Ponctoan Spear

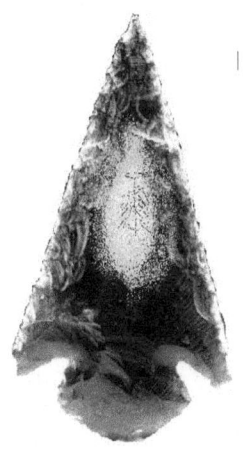

Ponctoan Spear
Title ID: 3922352
ISBN-13: 978-1478148982
**ISBN-10:
1478148985**

Dedications go out to my wife Beth

~to Betty Walters (my Mother)~

~~and my grandson Alex~~

PONCTOAN SPEAR

South Florida: 1530AD and Indian tribal wars are raging. The Ponctoan Tribe is teetering on extinction. An ancient relic is passed on to a trusted Ponctoan Warrior by the tribal medicine chief. With it, he learns powerful lessons. Lessons that will level the playing field in war; and that begins with a time travel odyssey, that starts in Davistown Pennsylvania in 1966. With him, he takes the tribal war spear of his people. It is his only ticket through space and time. However, there are those who will do anything to acquire it and the young warrior must reach into his ancient training of war to survive the future!

Table of Contents

Ponctoan Spear

Prologue

500 million years ago when arthropods ruled the depths of the oceans, the lure of shallow expectations became a new frontier. Coupled with warmer water and copious outcropping of algae and plankton made the journey an enticing one. Millions of these small creatures known as Trilobites took to shallower areas where getting close to land seemed to be the next gigantic leap to managing a piece of real estate in future development plans. However, for some, those plans would have to come much later; at least, 200 million years in the future.

Up from the depths came a swarm of Trilobites with the same intensity as a swarm of locusts across the sands of Egypt; and they came equipped with evolution-ally enhanced features unmatched by their old predecessors. Such enhancements included eyes and hearing. These newly enhanced features would serve to become very handy for their new shallow-water environment.

All throughout the millions of Trilobites in the swarm, one such specimen had the developing stages of both sight and hearing. One such little one had all the enzymes running throughout her body to make the link to this step of evolution complete. And now it was; she began her first season of procreation. Inside her brood pouch her eggs held tight. And in this warmer environment, there was excitement about her. She somehow must have known the warmth was good for her babies, known at this stage as simply larvae. And now, apart from all others in the swarm, she was the mother of the new generation of arthropods. Perhaps the first mother to deliver eggs into larvae stages in shallow waters. And so Trilo was her name, for this period of time was all hers.

Trilo went enthusiastically forward with what she knew best; scouring the sandy bottom for minuscule sea life, both plant and animal and coping with the tiny larvae wiggling in her brood pouch, keeping her awake at night. Trilo would ball herself up to protect herself and even slither under the sand when she slept. Keeping her babies safe from danger were her main drive in life. Nature would handle the rest.

And in nature Trilo went through her scheduled life cycles. As she aged she shed her exoskeleton many times, each time adding with it another segment to her body. Trilo was a young one when she began giving birth to new Trilobites. In a sense, she was still growing, though now, barely larger than a human tear. Her vision began to suffer as each shedding took a layer of the lens from her eyes, but her hearing was still quite keen.

Far far away Trilo heard the rumbling. They came from parts far away and far from the water's edge. And that was okay for no fear came from out of the water. However, today it was different in so many exciting ways. There was rumbling sounds from the depths of the sea. From a place, she can only recall but barely remembers. Suddenly the shallows rose and then fell and with it, she felt suspended in an unfamiliar manner. In all her senses she immediately realizes she has ability to experience yet another. Could it be a sense of smell? Trilo is not afraid. She has had the call of nature summon her from the depths. Today she will embark on another. The force of nature is to be trusted and she is tossed from the shallows atop a mound of soft dry sandy beach. Softly she flutters her appendages taking in the strange sensations that overcame her.

Instantly, the outcropping of brush and jungle is trampled out with a humongous wall of red churning lava. Three super volcanoes have reared their anger on the land and sea. And the ocean literally begins to recede under thick blankets of steam and smoke. Trilo senses the odor and is blessed for no trilobite has evolved to acquire a sense of smell. Even so, in her crude bio-chemical makeup, the smoke and steam were so intense that she literally sampled the absorbed mayhem through the tissues of her gills. Though she was in mortal danger Trilo was incensed by it all and took in each moment as it came.

And came it did. A giant wave pounded down upon the little mound of sand where Trilo lay. The force drove Trilo nearly a meter into the sand. And as it did her little teardrop-shaped body vaporized into nothing more than a gas as temperatures hitting her were an instant 7 thousand degrees Fahrenheit.

Volcanoes raged and ravaged the area for nearly a century thereafter. The effects were only temporary when compared to what effects Trilo left behind. Her remnants remained encased in a volcanic glass known as obsidian. The gaseous bubble that was a remnant of all Trilo was throughout her life became stationary in the glass as it cooled. Trilo, now suspended forever a few hundred millimeters below the surface of the glass itself.

Chapter 1:
The Americas

14th Century Florida

His name was Huega. And he was a Warrior and gatherer to his people, the tribe of the Ponctoan Indian. Nomadic as Indians of his own time, he has seen most of the other tribes creating settlements for themselves. As a result, political influence has made life hard for those of free spirit. People of who hunt warm blooded game and those who do not crop the land for food. Even trading with others of different cultures have become suspect for trouble. Originally, the Ponctoan roamed about the circular shores and swamplands of Mayaimi (now known as Lake Okeechobee). As one species of animal moved in others would move out. Such would be the fox after the rabbit, the cat after the fox. Hog after the snake and the cycle continued in a circle around the great Mayaimi. And then the hunters of man moved into the circle; the Calusa warrior. This tribesman was the worst, a heart like that of an animal. And he lived and took nourishment from 'ghost meat'. Their numbers multiplied like field mice and they lived atop the dumping of the food seashells they cast aside. With that the Calusa had a saying, 'He who walks on shells walks silent in the forest'. It meant simply 'toughened soles'. The Calusa could literally walk over sharpened tools without a single cut! And yes, they were dangerously silent. Unlike the Ponctoan people who became nearly invisible in the forest, often feeling at home in the trees, the people of the Calusa tribe were coastal dwellers. The Ponctoan lived quietly and slowly. They observed and marked important objects, places and property. They were masters at setting snares for both animal and man. So it was for the first time in the history of the Ponctoan, to build a settlement on the southern shore of the Peace River, thus giving up their nomadic ways forever. A new problem arose when unrest among neighbouring tribes, created by the Calusa Indian nation, made Huega feel the settlement was a temporary way of making a defiant stand.

Trading between other friendly tribes has all but becomes non-existent in the interior regions of Florida. It has become obvious to Huega that the Tequesta had migrated to the southeast near the white shoals of the Biscayne seas and the Jaega tribe as well. Huega now could not recall the last time he ran into a member of the Ais Tribe. They were people of cautious hospitality. The level of their trust for you increased with each gesture of kindness and thoughtfulness that you extended them. Huega shuttered. The Ais may have moved to the badlands of the north. The regions to the north were becoming more and more in control of the white man. He sighed. It seemed the world was going to hell and the only refuge left to go will be to the south. Soon it will be time for the Ponctoan to migrate south. Calusa territory was growing and there would be no more talks and no more compromises.

The sun did not yet set and was a full hand above the grassland behind him. With his condition, he desperately needed to find a place to bed down. Huega was two days off from his tribal settlement. His village resided on the southern banks of the Peace River near Mayaimi. In the journey before him, the forest bulrushes were giving way to thickets of small mangrove; and the shadow of the great eagle soared overhead. There was a brief moment when the wind shifted out of the east, he could smell the sea. And now, out of that, the trace of a stranger? The eagle was cautiously circling suspiciously as if her nest was threatened.

Huega froze. His nostrils flared. Slowly, he inhaled, tuning his head right then left. There it was again ... Sweat! It was the foul sweat of a mollusk eater. Eater of ghost meat; and in a bit of a turn left he noticed the sawgrass lean left of center in a spot not more than ten paces away. Huega slipped to a crouch, hiding himself in the tall grass. It was now a game of wait. The secret was in never doing night battle in an area strange to them. Snares and traps were part of the battle gear and leading the enemy to those traps and snares were easy.

Nearly an hour passed. With each warrior bobbing up and down to get a look at one another, it became obvious to each, the folly in this tactic. Twice they had seen one another and twice they quickly concealed themselves in the grass.

The Calusa warrior slowly stood up and declared, "Ponctoan show yourself and die like a warrior!"

Huega remain concealed in the sawgrass, "I prefer to die under the stars."

"I will not wait. For why should I not kill you now so that I can spend the rest of the night with your wife and your daughter in bed?"

Huega held back his anger, "Because I know you would rather spend the night with a goat tied to a tree instead!"

There was a long period of silence before the Calusa warrior spoke again. "You do not know who I am. I am a great warrior with many enemy souls under my feet."

"You are a nobody wishing to bring back a warrior trophy. You kneel and kiss the hands of your high and mighty."

The Calusa warrior snorted in anger, "You are a NOBODY!"

"You knew that I had a wife and daughter," Huega exclaimed as he rose up from the sawgrass to face his foe. Huega whirled his head from shoulder to shoulder, looking for more Calusa warriors. However, it was plain to see there were none. This warrior was seeking to prove himself worthy of higher influence in his tribe and did not want to share in the manhunt. Even so, he seems to have forgotten one important thing. "So now that I know you will drag me back as a trophy of your warrior skills, who will drag you back to be buried after I kill you ... since you did not bring with you anyone to tend to your funeral arrangements."

Waving his knife around, Huega continued nonchalantly, "I think you failed to consider this; the fact that this is a really bad day for you."

The Calusa warrior's brow furrowed deeply as if the thought brought up uncertainty, "Bad day? What do you mean?"

"You failed to sneak up on me during the daytime. What makes you think you will have a chance once the sun drops below the grass?" This was Huega's best bluff. It had to work. He was recovering from a snake bite on his right calf. The medicine was strong and he felt feeble much sooner than normal. Even though it has been eight weeks, the medicine and the after-effects of the venom makes one a bit dopey for twice longer than this. Once the sun goes down, Huega will pass out, no matter what danger surrounds him. Already his eyelids felt heavy. He should be settled in bed by now, but if he is to survive to see the sun rise, he will have to act immediately and put this brute to bed first. "Turn for home now and live. Else you wake up in the morning a snapping turtle!" Huega was knowledgeable about the Calusa beliefs. The Calusa believed in reincarnation. In death, their souls moved into the level of an animal. Once that animal passed on; the soul would go down the next level and so forth. The Calusa hung his head. It was true. Not even the snapping turtle with his head severed from his body would hold fast to his prey in his jaws after the sun goes down. "May there be another day!" he exclaimed as he turned and went about his way home.

"Fair enough," muttered Huega. He was passing out and could not make another step. He would sleep here for the night, in the tall sawgrass. The last thing Huega felt as he dropped suddenly to his knees was nightfall. He literally felt as if the weightiness of the night collapsed around him with a hefty thud. His side hurt for a moment but then the pain dulled and he slipped into deep sleep.

*****

Misty beams of early morning sun came filtering through the grass and Huega awoke feeling the pain of a headache. He was wet with dew and knew his headache was inevitable for laying out all night low in the grass. Drawing a deep breath from the sigh of a yawn, his side hurt as if he had sustained a punch or a kick in the ribs. Other places of injury such as the snake bite were checked and all looked well so Huega launched himself to his feet.

The Calusa Indian warrior did not go far. Much to Huega's surprise lying next to the right of where he lay for the night laid the warrior. He was face-up and staring blankly up at the sky. There was dried blood from his nose and he appeared cold and dead. Huega knelt next to the warrior. Clenching the man's head by the thick mane of his raven-black hair, Huega checked him for a broken neck. It occurred to Huega that the feeling of this warrior charging him in a surprise attack was what accounted for the accidental kick in his side. From there, the warrior tripped over Huega when he unexpectedly dropped to his knees and then tumbled uncontrollably onto his back. The way this warrior laid made an impression in Huega's mind. He must have died instantly!

No broken neck, but when Huega lifted the warrior's head, he became surprised at how this warrior's head fell apart so easily. "I see you chose the life of a snapping turtle. May your new world be a good one."

Huega drug the warrior a short distance and then returned with the question in his mind as to what shattered this man's head. And what he found both shocked him and confused him. The discovery; a rare one, for the most part rocks were very rare in this area of Florida. Even so, here was the most unusual rock Huega ever seen. He pulled more of the grass away to reveal a smooth black rock with a raised ridge about an inch high. This, he was certain split the warrior's skull like the edge of an axe. Most of this day was now spent washing this mysterious rock and of course feeding the alligators one 'has been' Calusa Warrior. It took over ten hours tracing out the size of this stone. And most of the next day getting it dug up. Once the main part of the stone was free from the sandy clutches of the earth, Huega discovered yet more left behind from the original stone. Parts were fractured away leaving behind glass-like shards. So razor-sharp were these pieces they cut through his deer hide legging with ease! It was a mere test cut, but Huega was so impressed he cut a few more slits in his leggings.

Now the real work began as Huega set in motion a plan to get the stone home to the tribal settlement. There was no question he'd tote an 85-pound stone by himself, so he went to work building a travois suitable for pulling by human. When he was finished, it resembled a crude wheelbarrow minus the wheel. He then loaded up the stone, tied it down with leather strapping from his britches and began dragging the stone home. The two-day journey took nearly three days to complete. And when he arrived and cut the leather strapping, the stone rolled down and settled by the feet of a proud Ponctoan Warrior. He had so much to tell his people but most importantly his Chief and Doctor. Although it was getting near dark and dusk had settled in and damn that snake serum. It still ran through his veins.

Huega dropped to his knees and before he landed on his face, the Chief caught him and had him taken to his Ceremonial House to spend the night in comfort.

Huega awoke to the radiant face of his little girl. She stood patiently holding a clutch of beautiful snapdragon flowers. "Father," she said softly. "I missed you."

A big smile broke across a weary mornings' awakening. Every force in Huega ached, but here stood, the best medicine a Father could have. "Why it's my little Chickpea!" He reached around to pull her close when she dropped her flowers and slid up onto the bed beside him.

"Tell me father, of your journey to find the black stone!"

"Ah, so already you have seen this stone," Huega said. He slid himself up to sit beside his daughter. Little Chickpea was only five seasons old yet she was smart for one, twice that age. "It was many days into my journey hunting the giant eagle. An eagle the size of this entire house! But it was not the eagle I hunted for. It was for her secret egg. Many warriors have hunted the secret egg for its gifts it brings the people of the tribe."

"What kind of gifts Father?"

"It brings tools for the elders to use to hunt with and to make wonderful things."

"What kind of wonderful things Father?"

"Wonderful things like the pretty dress you wear to the pretty little dolls mother has made you!"

Chickpea quivered with excitement. She clapped her hands. "Tell me-tell me more Father. How did you get the Giant Eagle to give you her secret egg?"

"You see the Giant Eagle would happily give me the secret egg on condition that I destroy the evil Buzzard Bird who kept her from her nest ..."

"Oh Father! How did you destroy the evil Buzzard Bird?"

"I pulled a feather from his tail when he wasn't looking. And when he squawked at me, I shoved the feather down his throat until he choked. By morning he turned into a snapping turtle! And you know what?"

"What," Chickpea replied.

"Everyone knows turtles aren't a threat to eagles. Especially the Giant Eagle! And you know why?"

Chickpea didn't hesitate a moment. The answer was clear. "Because turtles can't climb TREEES!"

Standing in the doorway was Chickpea's mother, Luega and wife to Huega. Seeing her standing there made him realize where little Chickpea got her beauty. Today, Huega was a proud man. A very sore man too!

Chapter 2:

Of Parallel Discoveries

Davistown Pennsylvania January 1966

It was the poorest side of town and the sidewalks were uneven. Walking in heavy snow made it a bit tricky for Professor Wheeler to anticipate his next step. For a man who was making a recovery from polio, he walked quite well when the weather suited him. Alan Wheeler was only nine years old at the time polio struck and now at the age of 26 his stride, became a challenge in such conditions. He keeps a cane close at hand, but given better conditions, walks well enough without it.

Wheeler was on an urban mission to the Mac Charles Home for Boys. It would be his first visit there since meeting a young man who claimed to live there. It was an odd sort of meeting. The man seemed precisely that, a juvenile. At first, it seemed to Alan; this man was only a boy. However, something about him told otherwise. The boy carried himself maturely in a way that baffled Dr Wheeler. Could it be there was more to what he had told Dr Wheeler a few days before? Alan Wheeler stopped a moment and shook his head.

I am an astrophysicist, not an MD. He was only thinking it, but yet he could hear his voice echo in the frozen branches of the trees overhead. *I am,* once more he thought, *only paying a visit to hear this young man tell me once again, what he stated ... a number of days before.*

Alan's mind wandered to that first meeting. It was as if this young man had never seen a PhD or Doctor. Or did he? Over and over he had asked Professor Wheeler if he was a Doctor or Chief. Alan explained that he was faculty and not an administrator, a Chief like a Dean, or head of anything here at Davistown University; a Doctor of Astronomy and Quantum physics. Recently Doctor Wheeler made strong arguments on both sides of the science community, namely that of Relativity and Quantum physics. Alan was always one who kept an open mind and would listen to everyone with an opinion regardless of what walk of life they came from.

Slowly and respectful of Alan's condition, the young man walked alongside Doctor Wheeler as they both headed to Wheeler's study.

"You were wounded in battle?" the young man asked.

Professor Wheeler hesitated for a moment. It was clear the young man would tag-along asking question after question of him, but why this one?

"If you are asking if I am a draft-dodger the answer is no," he replied tersely. "So if you'll excuse me, I have a lecture to make in less than an hour and need to review some notes."

The grip he put on Alan's wrist was in texture and pressure not of a young boy. His eyes pierced through Alan's impatient stammer, as if to be probing Alan's eyes for a lost soul or spirit. The young man's eyes were as ebony and stopped the professor in his gait. They stood motionless for a moment. Then slowly the young man released his grip as if melting back into the adolescence of a scolded boy. He then pointed to his chest. "I am here," the young man said. "And you are there." His hands came together as he raised them to the ceiling. "And in the light of the next moment we are far ahead of where we were."

Obviously, the young man was not an astrophysicist by any stretch. His words came slow and clumsy as if he had a native dialect similar to Hispanic perhaps. Nevertheless, this statement he made caught Professor Wheeler off guard. The idea suggested what might be another way of expressing Relativity. But of course, one thing was missing. Could it be the key that was missing? Or rather, a symbol of what, the young man held in theory?

Doctor Wheeler's heart leapt at the thought. He too had been secretly working on this very same theory!

"See here," Doctor Wheeler said. "E=MC square is pivotal to the relationship of mass and energy and ..."

"Darkness," the young man injected slowly.

"NO, Light. Now as I was saying," Doctor Wheeler continued.

"Light gives way to Darkness." The young man hesitates but only stares off as though the professor was not there. Then he slowly ends by saying, "and so gives way to time."

"I, I was … about to," Professor Wheeler responded slowly, "but what, if I understand you correctly, are you stating here? That there are such things as interstellar Black Holes?"

The existence of Black Holes had only been locked in theory but not realized as fact. It was his life's work; a mystery, Doctor Wheeler spent his entire career trying to solve. But then the young man just stared off again as if in deep thought.

"Excuse me," professor Pope said before passing by Alan Wheeler and the young man. "What the heck was it you wanted me to bring you back from Florida?"

Doctor Wheeler excused himself, turning away from young man to consult with the History Professor. "You wanted a mango right?" Doctor Pope asked.

"Listen," Doctor Wheeler says in a hushed tone, "this young man whom I am talking with; I can't explain it but something about him tells me he has some sort of celestial beliefs."

"Celestial …? Ah, okay. So?"

"Well yes. In my area of expertise, I believe he sees the mechanics of the universe in ahem, sort of the same way as I do. His model expresses some interest to me."

"Doesn't look like a mathematician to me."

"No Clark," Alan said. "I didn't mean a math model of his theories. I just thought you might tell me what nationality he looks like to you."

"Oh right. Right. Like you said celestial beliefs as akin to his nationality perhaps?" Professor Clark Pope just stood there grinning. Perched on a verge of laughter.

Professor Alan Wheeler stared back waiting for Clark's opinion.

"You're kidding right?" Clark asked.

"Just bring me back a Mango," Alan replied with a sigh looking backward just in time to see the door at the end of the corridor close shut. The young man was gone. He simply fled as the door from the other end of the hallway opened. And instinct for trouble Alan surmised, as a security cop accompanying another youth by the arm walked up. "Did you see another kid come through here?" asked the security cop.

"Did he look like a wild Indian?" asked Professor Pope?

"Yea and he'll be wild by the time I get through with him!" snarled the security cop. "They're the hoodlums from the orphanage on 23rd Street, out to rob vending machines! Caught this one with his arm stuck in the sandwich machine."

"Indian, you don't say," pondered Alan Wheeler out loud. "what sort of Indian?"

"So which way did the other one go you say?" asked the security cop again.

Doctor Pope spoke up, "Didn't get a close look, but I think of Appalachian stock ... based on the high forehead and large cheek bones."

Doctor Wheeler looked puzzled.

Doctor Pope shrugged his shoulders with a goofy grin on his face, "Well, I *am* an anthropologist too you know!"

"Sir, I asked which *way?*"

"Oh yes. Sorry ... ah, I think that way." Doctor Wheeler pointed.

"That's the same door I just came through! That's *impossible!*" exclaimed the security guard.

"Then why'd you ask?" Alan replied. Still chuckling, he turned to Clark to wish him a safe drive down to Florida.

"Mac Charles Home for Boys," Clark repeated slowly as if in deep thought.

"What's that Clark?" Alan asked.

"The home of course. You know the orphanage on 23rd Street!"

*****

"Since my husband Mac passed away a few years ago; we have had to investigate who we take in here at the home." Mrs Charles explained to Doctor Wheeler over coffee. "You see the boy you are interested in seeing came here as a referral with no name and no information as to his next of kin. He's only been here a little over a week you know. He was in trouble you see and had it not been for my husband's brother Sam; I would not have taken him in."

Mrs Charles was now about to tell Alan that the boy was no longer with them. "So it was about the incident at Davistown University yesterday?" Alan asked.

"Oh my yes," she replied. "I had a feeling about this young man. *And* that he was going to be trouble. A bad example to the other boys you know. He, being *dressed* that way and all." She went on to explain about all the deer hide leather and paint and the curious pouch of arrowheads he had with him.

"Was that the only thing that bothered you about him?"

"Well since you did ask," she said. "I looked at his incarceration report and on it was John Doe. Well we all know what *that* means. It means arrogance and trouble. He wouldn't talk you know. To anyone, not even me!" She hesitated for a moment and sipped her tea. "Like I said, my brother in law Sam had him working at the drug store and said he was a good worker and even saved his life from some hoodlums who wanted to rob him *again!*" Then clearing her throat she continued. "Well. When he arrived, I told him I was in charge and he laughed! He took my hand and called me mum and rubbed it and said Michea."

"And that angered you?" Alan asked softly seeing her eyes well up in tears.

"I don't know," she finally replied looking away briefly. "I just knew in that moment he was a good boy. And he meant well with his silence and all. And besides I had a brother who was obsessed with hunting for arrowheads when he was a kid."

"So you told him to leave, or he ran away I presume?"

Mrs Charles gasped a sigh of relief, "*Land sakes,* Doctor Wheeler! Do I look like someone who would do such a thing?"

"Of course not."

"He went to the men's shelter."

Doctor Wheeler shook his head a bit confused. "He went where? You mean Salvation Army of course, right?"

"Yes and not on his own. A man came by yesterday and picked him up and took him there, "

"Who was this man?"

"He was concerned over the trouble that went down at the University and although Michea was not caught or officially implicated in the trouble, he confessed he was there. Said something about knowing the man who came and took him to the men's shelter."

"The security officer I presume," Doctor Wheeler added.

"May I get you some more coffee?" smiled Mrs Charles.

"No no. I've been enough trouble." Doctor Wheeler replied.

She walked him to the door and the hat rack where his hat hung adorning a sculpted mirror. Maybe it was the light in the foyer, but it looked as if he had picked up a few greyer hairs at his temples. Taking his cane in hand, he carefully approached the porch steps one step at a time. Alan would give the men's shelter a call from his lab.

*****

It was 5:30 PM when Alan reached his office. A take-out order of fried chicken and a fountain coke with biscuits in a bag as he walked up to the door of his lab/office. It had been a busy day already and he was certain a long night was ahead of him. He had so much to do, but he was hungry and settled behind his desk with his chicken dinner. The last bit of light was seeping through the blinds. He sat sipping his drink knowing he needed to turn on the lights to read, but put off getting up from his chair right away as he felt a bit drowsy.

At 7:20 PM, a noise startled Alan. Groggily he wondered if the noise was outside his door in the hallway ... or in his lab. Once getting his thoughts together he realized how well these classrooms and labs were soundproofed. One could set-off a string of firecrackers in the hallway and unless it was right outside your door, you wouldn't hear them! He rubbed the sleepiness from his face with open palms; stood up from his desk to face the door where the light switch was near. Light from the hallway spilled into the room through the small rectangular glass window in the door. For a moment, it seemed so far away. More distant than usual and the window light flickered once. It was as if something or someone had just passed by the little window. The scraping noise Alan heard just moment ago was indeed inside the room. He was certain of that now.

Alan could hear his own breathing and the energy to move, instead, seized him solid like a statue. His heart pounded. The door was far away and somehow he felt going for the door would be a mistake. But then maybe, he was just imagining this. He gently leaned against his desk, drawing a steady breath to relax himself. Nothing like a deep breath to help calm you down he thought. He held that breath and for the moment, it seemed everything was all right. Except now that he was not breathing, he could hear the breathing of someone else instead.

And that someone was in the room!

"Who's there?" Alan called out.

Not far off the left ear, Alan heard a voice answer him. "I am here. I am your friend. Awaken the lights, so we will see well."

From the voice, Alan knew it was Michea. Even before reaching the light switch, Alan knew there was something Michea urgently wanted from him. However, what it was didn't take priority at the moment. Only that Michea explained how he got past the watchful eye of the faculty and staff. Most of all, how he got past his locked Laboratory door!

"I only just arrived," he said slowly, "and in the darkness of the night I am here."

Before Alan could ask how he got around a locked door, Michea pointed up to the ceiling. There, a misplaced ceiling tile looked a bit scraped-up with parts of it chipped off and laying on a lab bench.

So this was the scraping noise Alan heard just after having woken from his nap. It was an unbelievable feat. Having vaulted to the roof of the college, dropping into the ceiling and air-handler, Michea tracked his way directly to where Alan sat in his lab. Of course if he was a psychopath, Alan would have already been dead. Damn the idea of that, Alan reflected in thought. Instead, there stood a tattered looking Indian with tangled black hair. He was breathing heavily, his eyelids half open and looking down at the floor.

"So you come in peace," Alan exclaimed. "Sorry that was pretty corny I guess."

Saying nothing, Michea just took a step backward and awkwardly slid down with his back against the wall, crumbling to sit, head propped in his arms. Alan asked if he was alright and Michea slowly lifted his head.

"You are a great doctor," he said.

And with that, Alan approached him to get a better look at this young man. He didn't seem well. Michea seemed exhausted. "Why would you say that I am a great doctor?" Alan asked.

"It is known," he replied.

"It is known?" Alan asked.

"... You know me," Michea repeated, his voice faint.

Alan thought back to his visit this afternoon. It was about where he learned of this young man's name. It wasn't John Doe it was Michea. "Oh yes, the lady at the orphanage," Alan began. However, he stopped short of explaining how he found out Michea's name. He stopped short because this young man was burning up with what seemed to be fever. Alan's hand never touched him. He just held the back of his hand near Michea's forehead. From an inch away Alan could see Michea was in grave danger. Sickness fell upon Michea fast. By the time Alan searched the phone book for emergency numbers, Michea was talking out of his head.

Chapter 3:

South Florida 1523 AD

Lessons and Discoveries.

At the age of ten, Michea learned to talk to the birds and they seemed to talk back at times. Warm summer days spent lying about silent in the forest watching the antics of squirrel and bird, taught Michea much. Even so, it was his great elders; the Chief and Doctor of the tribe that taught him how to hunt and become invisible to others who sought to harm him. And there were many who wanted to destroy him and his people, namely that being the Calusa tribesman and the white man. And of it all, none made sense. Only hunt and be hunted; but then, that is what made life interesting and charming for a young boy. Watching all the other older boys step up into a higher class of standing in the tribe, made Michea's spirit sing for his time to come! Most would become honored warriors. Such honors included a chance to pick a wife among the women of ones' choice. Once a warrior takes a wife, he renames her as his, with a new name different than her maiden name and yet, nearly the same as his own. Even at the tender age of seven seasons, Michea had his eye on

Chickpea as his future wife. Her name will be Lichea, he knew. And, he would name her proudly for himself one day.

Since being shot with an arrow in the back several weeks ago, it was nice to be able to get around and visit his friends of the forest. The arrow was a lesson learned. Sometimes, there is a price to be paid for mischievous behavior. Even so, Michea had a long time to think about his injury while he was recuperating from it. Had an experienced warrior done what he had done, he would have been decorated and honored. And not chastised the way Michea had been when he barely returned to the settlement alive with his prized trophy. After all, it takes a brave warrior to have snuck into a Calusa Indian Temple and steal a prized statue of their god. He was wrong, for he learned, a Ponctoan warrior fights only for honor and survival.

Today there is much happiness and an air of excitement in the settlement. The Chief's Warrior son Huega has come home with a mysterious prize. A large egg shaped black stone. Michea has never seen anything like this before. As it is certain to him now, no one else in the tribe has either! A few small hands of the children gather to touch the stone. Some look close to see their faces reflected in the stone. The tribal Doctor has earlier declared the stone safe. Soon it is whispered, the medicine Doctor will bless the stone as sacred to the Ponctoan Tribe. He will declare to the people what this stone will become. Will it become sacred? Or the other hand, will it become a source or object for making tools and arrow tips.

As sacred, it will not be fractured-up for tools, knives, spear, and arrow tips. Instead, it will be polished and adorned in fine cloth, gold twining and other precious adornments.

The medicine Doctor was very wise. Many came to him for decisions of consequence. No one ever challenged the Doctor's decisions. However, anyone could present evidence to persuade his decision in a direction that made for good common sense. And that is exactly what Huega did that morning. It made perfect common sense that the stone was created by the sacred spirits of the land as a utility to be shared among the people. Huega had the evidence in a deerskin pouch that he kept with himself at all times. So to demonstrate his evidence that the stone be used in tooling and arrow making, he called upon the Doctor. The Doctor in turn, called upon the people to show presence and witness to Huega's evidence.

In the early dawn hours that day, everyone gathered before the arch of the ceremonial house. Large blocks of cypress wood served as a sacred alter with the Doctor standing before it. Addressing the people, the Doctor declared Huega, a presenter of evidence. Everyone in the tribal village was present. And all villagers approached the alter, to get a look at what Huega lay on the alter top before the eyes of Ciabee. Ciabee was the name of the Doctor and he nodded his head when he saw a shard of the black stone being placed neatly beside a shock of bamboo and a dead dragonfly. One by one Ciabee lifted the evidence up to be examined by him and to the villagers.

After setting the last piece of evidence down on the alter, Ciabee said, "Demonstrate your evidence Huega, so the people shall see your argument."

Huega lifted the shard of obsidian stone in his right hand and the bamboo in his left hand. Slowly and with great ease, he cut a slice of bamboo to an amazing thickness never before witnessed in a cutting tool before now. The sound of amazement echoed in the circular clearing of the tribal settlement. Whispers grew louder until the Doctor raised his arms to hush the crowd. Huega was not done. And when the focus of the villagers shifted back to the demonstration, Huega showed the dead dragonfly. Everyone knew how light and fragile a dragonfly was, especially the wings. You could barely touch them without ripping them. And yet, Huega's obsidian shard sliced them cleanly without so much as a tiny bit of tearing! Ciabee concealed his delight with a frown and a nod. He announced his decision would be made after the feast. Everything and everyone went silent. Ciabee returned to his house to think a while. The silence in the village was suddenly broken when an old woman could be heard saying, "There is to be a feast?" Another could be heard saying, "Summon the hunters and gatherers!"

"Yes!" cried another. "We must prepare for a glorious dinner and already the day is growing short!"

Michea's heart soared! He knew of a place where a covey of dove held a habitat. And equally where turkey loved to browse in the trampled grasses left behind by wild pig! He must get his bow and quiver of arrows at once!

Disappointingly, however, at the door of his hut, his mother stood, her arms crossed. This was in protest of his hunt. Mothers certainly can read the minds of young hunters. She felt he was not well enough for this kind of action. She spun him around peeled up the back of his blouse.

"Mother please," protested Michea. "I am not a child!"

"Hold still. And yes you are, *my* child!"

Her hand pressed against his healing wound. Around the tips of her fingers, she looked for oozing. A sign the wound was still healing and raw. So far, so good, she sighed.

"Okay my child," she said. "I expect you home long before sunset."

Michea knew all too well a request for such is one he must keep, otherwise suffer the embarrassment of being escorted home by a tribal warrior who'd be sent out to track him and bring him back. Another lesson Michea learned and one he intended to live by.

"I will Mother," he promised. "I will drag home a boar pig big enough to feed many!"

Michea gathered his bow and quiver of arrows. He was excited. This was the first time in several weeks since being injured that he had the strength to pull back the arrow on his bow. His mother watched cautiously as he tested his strength on the bow pull. She nodded with a smile. "Okay my little man," she said. "And remember ..."

"Yes Mother," Michea cut in. "before sunset."

"And birds only," Mother yelled as he was halfway across the village circle headed for the jungle.

It seemed no sooner had she yelled that a thought crossed her mind. She scrambled over to Michea's bed where he hid his stash of arrows. She searched them all and all that was left behind were arrows with bird-tipped arrowheads. The broad tipped few that he had were taken with him. At once, she dashed for the doorway of the hut. She bit her lip. Michea had already disappeared into the jungle.

She shook her head. A smile crossed her face. Michea was growing up so fast.

Chapter 4:

The Feast

Michea knew all the local game trails and knew the habits of the wild hog. Since it was still early in the morning, he knew he had time to get positioned for the kill. After that getting the kill home may take a good hour if he was successful. He'd have to make his kill before noon. Otherwise there would be waste in having to quarter-out the kill so that it could be cooked in time for the evening's festivities.

Nestled in his tree stand, Michea waited and waited. Nearly an hour passed and he became anxious about his decision to be here and not hunting pheasant instead. Unfortunately, he had only one bird tipped arrow to hunt with. Then all at once, he heard a crackle of twigs under the hocks of a brood of pigs. His heart was once again lifted as he silently took a survey of the hogs below him. Two males and one very young sow. The sow, maybe 20 to 25 pounds, was a good choice he thought. Luck was on his side as there was no mother around. But then at this sow's age the bond already was broken. If it hadn't, Michea would have passed on this little one. Taking her would be dangerous as the mother would hang around for hours maybe days waiting to take you down the first attempt she got! That was an old lesson Michea had learned earlier with his Father before he was killed in a Calusa ambush two years before. He had many fond memories of his Father. The best being the time his Father told him he was proud of his son Michea! They had not hunted much before that. Maybe a year, but this was a lesson he had remembered well. Now, it would be lessons he learned on his own. Getting an arrow lodged in his back was one of those painful lessons for which he would get his revenge one day. Nevertheless, for now, dinner was crowding next to an enormous snorting boar hog. The thought of taking this colossal hog was just a fleeting thought, for Michea was quite sensible for his young age. Taking down the adolescent sow may not concern this group of scavengers. They may just run from fear and not return for a while. He could drop down and bleed her out while dragging her home.

An overhead shot was difficult and would require the young sow to be off slightly to Michea's left. Patience was the key here. The arrow must enter between the neck and the shoulder bone to be effective. Any deviation and the hog will take leave of the scene and eventually die elsewhere. Being off more than that and the hog will suffer long days of healing and pain. Much the way Michea had with his injury. He did not wish this on an innocent animal. He measured his breath and slowly slid an arrow from his quiver. He kept his movements short and close by his silhouette so as not to alarm the pigs. The arrow with a broad tip made from a meticulously crafted arrowhead of shell, slid carefully into position along Michea's bow string. His eye trained along the top leading edge of the arrow. Pretending his eye to be the sun settling behind the horizon, his eye sank to the entire top surface of the arrow shaft. Just as he had been trained when he was a child, *soon the sun will sit on his victim,* he thought with pride. It was symbolic of course, but this was the Ponctoan way.

But then it dawned on him! The arrow was wrong! It was a battle arrow, not a game arrow! Game arrows, the arrowheads were vertical. It is because animal ribs are upstanding and that will ensure the arrow will penetrate the ribcage. Human ribs are horizontal and thus ensured proper penetration and a certain kill.

Michea would have to compensate by holding the bow in an awkward way. And as he did while the time was true, he pulled hard on the bow. He could feel the sting of the string on his right ear as he released the arrow with a snap.

The little pig never sang out in pain. She just folded silently and fell immediately as the arrow found its mark. Michea took in a gasp of air. He had been holding his breath for a long time now. His heart danced and he knew he'd make it back with this little one in plenty of time for the feast! His attention fled to the sting of his right ear. His hand went up to check his ear. A small trace of blood was seen on his hand as he swept across it. The bowstring clipped his ear. Michea was proud. Even though he was not a true Warrior yet, he had gained the mark of a true hunter!

Huega was the first hunter back to the settlement. Proudly, he presented his kill to the women-folk who were eager to get things started. This was big news around here and it would spread among his people. All the young hunters his age returned with small game. They brought mostly birds, a muskrat and a mature otter. The adult Warrior/Hunters contributed later with a number of wild turkeys and an 85 pound male warthog. Women-folk gathered berries, fruit; palmetto hearts to make swamp cabbage and several baskets of fresh sweet root potatoes and yams.

The entire village came alive and everyone was assigned a job. Most women worked in groups. And the chatter of gossip could be heard while strolling through the village. The men focused on skinning the kills and plucking birds and cutting meat. For the fires that marked the center of the village, were being struck and fueled with kindling and large chunks of dried wood. Turkeys were spiked on a spit and young hunters made to attend them while swapping stories.

Michea was not handed a job. His pig was being roasted over a bed of coals topped with green palmetto fronds. The pig itself was wrapped in broad leaf banana fronds, then set into the coal pit, covered with soil and allowed to slow cook for four or five hours. So Michea walked about the village with his chest pumped and his head held high. The way he figured it, most of the people will prefer tender young pork over a gangly old warthog! Similarly, the water-rascals like the otter and muskrat; tasty in the winter, sinewy in the summer.

The tribal Doctor would preside over the evening's feast, with special offerings. A decree was the topic of celebration tonight. Nevertheless, it was the matters of the elders to discuss such things. Things that mattered little to Michea, since he was too young for this sort of talk. Later he figured he would get the idea of what all this adult talking was about. For now, hunting and eating were something that Michea could relate to. It ran in his blood. He had a knack for hunting and trapping. That was what interested him the most right now. And when the evening passed, he ate robustly and slept well. His pork dinner was an enormous success. It disappeared and best of all, he saw the Chief relish it with a smile and a big nod that went his way. Michea's standing in the tribe was well established as a worthy hunter. And rightly before he reached his tenth season! His mother was quite proud of him too and often called him her little 'wild one'. She worried that he would pass to the next world like his father, because he was so fearless and brave. And like all Mothers, she kept her fears hidden and did all she could to promote her son's interests. As for the interests of the tribal society, it was decreed the stone would serve as utility. It's special worthiness was in the way it could be made razor sharp. Among the many thousands of game and war arrowheads this stone was expected to yield, many knives and scrapers could be added to that list.

Many of the womenfolk of the tribe were stone artisans whose job was to produce tools and arrowheads. Even so, tools were fashioned from mostly clam and oyster shell. Never before had obsidian been encountered much less recognized as a medium for making tools. Obsidian is fragile as glass. Only an experienced toolmaker can develop the best way to create these marvelous new kinds of tools. As it turned out Michea's mother and another woman was chosen to make one knife and one spear point each. The objects would then be presented for testing by the Chief and Doctor.

Michea's mother was excited to have this honour and spent all morning the next day sorting through her basket of chipping and carving tools. An assortment of hammers and chippers made of deer antler were selected for the job. Next, the call went out from the ceremonial house to present an obsidian shard to each of the two women. They would return tomorrow at about the same time to present their finished tools. The Doctor, in his wisdom would decide who found the best approach to crafting these tools. Together these two artists could collaborate and teach the others to make superior tools. For now, the two women would be denied visits from friends and work at each one's hut in solitude and thought. Luega sent her daughter Chickpea to stay with her grandmother until she was finished. She wished Michea's mother, Waukee luck with her shard of obsidian.

Waukee had no time to explain to Michea the shard of black stone she held in her hands. She needed a long time to study it, and to tap on it with one of her tiny hammers. She could see already, the value of this stone as well as the deadly sharpness they could have if crafted properly. She noticed how brittle sharpened edges could be and how they could easily shatter if cleaved on long angles. She wondered if one day to complete this challenge would be too short. Easy to see this stone held no similarity to anything she had used for tools before. There was plenty of material to make both a knife and a spear point if she mastered the material before too much went to waste. However, there was little room for error. And she began chipping small pieces from the edge of the obsidian shard ... when it happened.

The shard cracked diagonally with one piece falling to the floor where it landed on a small mahogany block of wood that she used as a cracking bench. When it struck the bench, it broke again. This broke the thickest part of the triangular remnant at its widest part. Waukee froze. She marveled how fragile this stone was. At the same time her heart sank it rose. Out of the cracking of this shard came the rough outline of her spear point. She would only need to angle one side enough to center the tip with the center lateral line. The line was merely in the eye of the artisan, but held function and form. From that she merely needs to apply the edge-chipping to form a reasonably sharp edge. After that she would concentrate on the remaining piece. It would be the very piece she would craft a sharp spear point from. Nevertheless, Waukee's excitement would fade fast when her eye caught a strange site.

On the piece of obsidian she had just blessed as the one for being the spear point, light had caught what appeared to be a bad flaw. She picked it up carefully examining what was clearly a bubble trapped in the center of the stone piece. It almost looked as if it were a pearl trapped inside the glass-like stone. She ran her finger over the slightly teardrop-shaped bubble. The surface concealed the bubble in a convex-like lens. On the opposite side of the stone, the bubble was just as clearly seen and under a convex-like surface as well. Right away this could only serve to weaken the spear point as this obviously would serve as a breaking point. *Or would it?*

She continued working with this piece. After all she needed the practice and if it should break or chip improperly, nothing would be lost as she deemed it to be defective anyway. It would not take long before Waukee found working with this obsidian stone was easier than with her former shell counterparts. Soon she was yielding perfect chippings. Before long, she held the spear point in her hand. She gasped at such beauty. It was flawless. Never had she fashioned a more elaborate spear point as this one! Even the teardrop-shaped bubble was centered perfectly with its pointed end on center-line with the tip. Even so, Waukee was being silly she knew. She had become lost in her work and in her fatigue; she became overly partial to her own work. It was time to get started on the other shard. She would craft a beautiful knife for the tribal Doctor.

Chapter 5:

A crack in time

Early the next morning, Waukee was crossing the village circle holding both tools in her hands. She paused for a moment to admire this spear point. The obsidian knife was beautiful. Far better than the scrapers and small knives made of sharpened conch shell.

Waukee was met at the ceremonial house by Luega. She had just finished showing her stone tools to the Chief. And now, Waukee placed hers before the watchful eye of the tribal Doctor. His hand immediately reached for the spear point. At first, he gripped it tightly in his hand as if getting a feel for its weight. And now he studied the knife. He held it nodding his head and frowning. It was a frown of disbelief that such a tool could have such a good weight advantage over its predecessors. Now came the test. The tribal Doctor lifted from the, alter an oak branch. He pulled the knife across the oak and a nice curl of wood led the edge with ease. Impressed with this fine knife, no one was paying attention to the Chief until the light dimmed within the interior of the Ceremonial House. The Chief stood in the doorway looking out to the village circle. His mass blocked most of the morning sun that filled the room. The tribal Doctor and both women now had their attention drawn on the Chief who held the spear point Waukee made to his eye. It was if he was seeing something inside the bubble suspended inside the spear point.

It was more than a casual look into the bubble than one's curiosity would lend. And Waukee became inquisitive about this and what the Chief thought he was seeing. Although before she could express her curiosity, the Chief wandered outside as if following something. As if a child with a butterfly net after a rogue butterfly, the Chief pursued what he could see through the lens of the spear point. Swatting at things no one else could see. Then it happened.

And as quick as it did, it was over. It ended with the chief crying out something from across the tribal circle. He dashed across the circle with a wild look in his eye, exclaiming how cold it was. His had gripped the spear point tightly as if he was afraid of dropping it.

No one saw how he left, only how he came back from a place just a few seconds away. According to the Chief who claimed he took a direct path back to the Ceremonial House, but ended up a distance away. However, it did not end there. He saw the entire day in the few seconds he was gone. And for the rest of the day he sat quietly by the door of his hut observing Déjà vu. The tribal Doctor silently observed the Chief from the doorway of his hut next door. Now and then the Chief's chest would rise and fall as if he was taking in a deep breath. In his hand, he clutched the spear point as if it was the most prized of all that was precious to him and the tribe. The Doctor joined company with the Chief, but the Chief wished to be alone. So the Doctor resumed his observation of the Chief from his door step.

Now and then the Chief raised the spear point in his hand and just looked admiringly at it. There was whispering in the tribal society. The Chief had not allowed this spear point to leave his sight. In fact, the Doctor couldn't say for sure whether he had even let it leave his hand. It was clear now. The Chief would make another declaration soon.

Chapter 6:

Davistown Pennsylvania January 1966

Alan Wheeler had the laboratory lights on and the phone book open. He frantically searched the book for emergency medical services. Ambulance services in particular.

Alan's heart pounded in fear for Michea's life. He had never seen anyone become sick so fast. Rivulets of sweat that had earlier poured down Michea's face and neck had dried and now he felt as if his temperature rose. He then closed his eyes and began to pitch his head side to side shouting words that Alan could not make sense of.

Alan had to shout over the phone. Michea began ranting. And in that rant his voice harmonized to make a tune for the ghostly words he chanted. His eyes opened and in those glaring eyes were the haunting mosaic of his past life. His eyes closed again and his voice went shallow and it quivered like the soft sounds of a dove.

"Get an ambulance out here, asap!" Alan was desperately pleading with the ambulance company operator. "Davistown University … yes, East Wing entrance, A or B. I will check you in on both intercoms. Please hurry!"

Alan quickly checked his request at the central office. And now that he alerted everyone in the East

Wing of the emergency, no time was wasted getting the emergency service paramedics on the scene.

Paramedics immediately cuffed Michea for a BP. "Let's stretch him for transport," one said to the other. "I can tell you his heart rate is high ... very high," said one paramedic to the other. "Temperature is 102.7 and holy cow, his blood pressure has hit the roof!"

"Ship 'him," snapped the other paramedic, "let's go!"

Alan nervously shuffled papers and then switched to winding his wristwatch. He was no good at this sort of thing. Although in the time it took for him to finish winding his watch, Michea was out the door on gurney. "Hey mister," yelled a paramedic over his should on his way out. "If you are with us let's go!"

It was natural for Alan to assume he should follow the ambulance in his car. But then the invitation to ride overcame him and he hurried in behind the paramedics, never once concerned about locking up his laboratory.

Rolling down the long hallway, the paramedic in the rear of the gurney kept looking back at Alan. Then finally, he asked, "You aren't kin to this fellow are you?"

Alan fumbled with his words, "ah, well, no."

"Oh," said the paramedic, "one of your students?"

This was all happening too fast. Alan was not sure how all this was going to work out for Michea. Obvious, this young man could not pay for his medical expenses. Agreeing to say yes that this was one of his students would be a lie. Then again, simply saying this man was merely an orphan or most reasonably, a homeless man, may boost his chances of getting help. Alan had never had to make life-or-death decisions for anyone before and this was just happening too fast.

"No," Alan began. "I mean yes … I mean this man is my friend is what I meant!"

"Okay," replied the paramedic, "you are going to be the voice of this one. We are going to need some information on him, so relax. He's in the best of hands right now."

They had already started an IV on Michea as well as strapped his hands down to the gurney. With all aboard and on the road, Alan could hear the radio contact with the emergency room physician who was instructing the paramedics to draw blood and get other vitals, including eye pupil dilation to light. And then the worst of all came. They had no sooner reached the circle drive of the emergency room slab, when Michea's heartbeat flat-lined. Alan could see the emergency room physician through the glass double doors. He was talking on a portable radio to the emergency unit paramedics just a few feet away.

"We have cardiac arrest."

Before Alan could turn his head, there at the loading door of the ambulance stood the emergency room doctor. One of the other paramedics who was inside the ambulance besides Alan asked him to jump out of the ambulance. "We need to give the doctor some space to work," he said. No one needed to tell

Alan this as he was already working his way out of the ambulance. Alan moved as fast as he was able, but the doctor was shoving in past him and literally popped Alan out onto the pavement. Falling the way Alan did, caused him to land on his side. He felt his shoulder snap and his head smacked the pavement.

All accounts that went on around Alan rapidly faded as he slid into unconsciousness.

Chapter 7:

Festival of the Ponctoan Spear

The day following the discovery of the spear's mysterious power, the Chief, the Doctor and Michea's mother held a private meeting in the Ceremonial House. The Chief told them of his experience with the spear point.

When looking through the bubble in the center of the spear point he claimed to see what looked like many tiny lights. Lights that grew bright then dimmed but really never disappeared. They just moved slowly about like fireflies in the night. And when he tried to study them up close, he said he discovered them to be little tiny light circles. He saw floating circles of light that surrounded a great darkness. The Chief drew a breath and paused in deep thought. His hands moved in a big circle that slowly came together in the center of his chest. The Medicine Doctor claimed, that the firefly itself was not an embodiment of light, but of darkness surrounded by light that poured into itself as if drinking it in. Upon closer inspection he could see this rim of intense light invert itself as it entered the great but tiny fleck of darkness. The Chief recalled his experience with great emphasis on how this light responded to his hand when viewed through the spear point. He said he could actually poke the small black holes with his finger and expand them a little. So it was a carefully measured step he took when he slowly lined the spear point with an eye on the black speck of a hole and then at the right moment, blindly thrust the spear point into the tiny celestial mark.

The Chief looked around him. The faces of both the Doctor and Waukee were in awe. It was then the Chief bravely continued.

On instinct the Chief claimed he made a slashing motion and literally ripped a gaping hole in the village landscape!

"It must have happened very fast," the Doctor said. "I saw you one moment at the doorway outside. The next moment you were running out of the forest from across the village circle!"

"It was then I jump through this rip in to the spirit world," The Chief slowly shook his head, "I do not know how this can be, but it was. And I was in this new world a long time before I jumped back."

"But you seemed to be there," The Doctor said pointing to the general area of the door steps. "Then you reappeared way out there. It was only an instant!" The Doctor just as confused as Waukee, as she expressed her astonishment as well. "It is so," she said. "you disappeared and then reappeared faster than the striking viper! What did you see throughout your travels in this other world?"

In his hands the Chief palmed the spear point. He looked thoughtfully at it and then stared out through the door of the Ceremonial House. "Follow me," he said to the two standing there in watchful anticipation.

Beside the doorstep he stood, motioning with the entire reach of his arm of a line that ran the length of the village ... and beyond. It was here he said that these tiny holes drifted about. Some slightly larger than others, while some consumed each other that came too close. Very carefully and with both hands, the Chief passed the spear point to the Doctor. "Observe," the Chief directed.

Holding the spear point close to his face, the Doctor lined-up his right eye on the center-line of the spear point until it crossed the teardrop-shaped bubble inside the obsidian. The Doctor said nothing. However, it was clear he was seeing what the Chief had seen the day before. He stepped forward and the Chief grabbed his elbow as if to steady his steps. The Doctor, with a free hand, reached with curious tenacity at the holes rimmed with bright pearls of light. The Chief felt the hair of the Doctors elbow and forearm bristle in anticipation. He could be heard saying, "They're beautiful!"

For Waukee, she saw nothing of what made the Doctor act as if a giddy child. This was something she had never witnessed before of this wise one. All the while she stood there she must have felt special for having created such a curious spear point. Now that it was obvious the Doctor was experiencing much the same thing as the Chief had the day before, what happened next was perhaps only expected by the Chief. After all Waukee never expected the Doctor to try to repeat the Chief's experience. She saw the

Doctor swing the spear point with the exuberance of one ripping the hide skin off an antelope. And following such, he disappeared. The Chief stood motionless for a moment, his mouth agape in awe.

There, standing in the center of the village clearing stood the Doctor, spear point in hand. The mane of his thick black hair covering his face as he looked down at the spear point he held. Looking up to the sky the Doctor broke into a song of praise. And then he slowly knelt still singing. He sang of how he just fell back from the stars to his home. The Chief knew the truth in the Doctors lyrics. And of the praise he sang of the spirit trapped inside the spear point. Thinking back to his own experience, the Chief concluded that Doctor must have slipped from the mystical precipice that ran the length of the village and beyond. That would have made his journey a short one as it was clear to the Chief, the Doctor did not journey down this walkway through time, but rather, stood in place until the walkway crumbled from below his feet.

The Chief could not understand why at first the walkway slowly crumbled away forcing one to advance their walk ahead of the deteriorating stones of the walkway. Because of this, there was no retracing one's steps back. All along the walkway was images of the village activities as they progressed throughout the day. Along the walkway beyond a slow walk, one could see the entire day from morning to sunset and the beginning of the next day. And so forth. The Chief travelled this short distance watching inquisitively the day and all activities as recorded by time. By the time the walkway had crumbled to where he stood, he had seen the entire day. Knowing he could not go back nor find his way back, he leaped out over the void of nothingness closest to where he could consider his entry-point into this place. So it was and he was certain, the Doctor's visit was shorter. Even so, the Doctor saw much. The day will unfold as predictable as yesterday to him. *He will be anxious to talk about his experience,* knew the Chief.

By now, many of the Ponctoan Indian villagers in the settlement gathered around. All were aware of today's meeting and now many were whispering among themselves. The Chief nudged the Doctor for his hand and the Doctor responded in kind. While the Chief escorted the Doctor back to the Ceremonial House the villagers cheered. While everyone knew of the special spear point Waukee had crafted for the Chief, most suspected it as spiritually important. However, of what the Doctor sang out, left many unanswered questions among the villagers. Later those questions would be answered but first there would be another feast. That would come tomorrow and it would be the Festival of the Ponctoan Spear.

With Michea trying his best to repeat his last hunt, he bagged a turkey and a wild rooster instead of a hog. Nevertheless, it was this festival that really captured Michea's imagination. This strange spiritual world was accessible through the eye of a spear point! While everyone was permitted to walk through the Ceremonial House to see this Spear Point up close, no one but the two elder leaders of the tribe were allowed to handle it.

Everyone except his Mother, thought Michea proudly! "What was it-like Mother?"

"It was like this one," Waukee said. She held up a small obsidian arrowhead that she had made just for him. What a surprise! Waukee had made her son one of the first obsidian arrowheads the tribe has seen.

"But Mother," Michea said with a gasp, "what about ..."

Waukee cut in, "The Chief wanted you to be privileged." Waukee knew it was an honour given to her, for she was given the honour of instructing others how it was done. Making arrowheads that is. The rest comes down to the arrow makers who generally were old men of the tribe. Sometimes Michea sat for hours watching arrowheads being bound to the shafts of willow wood and other various woods chosen for being true to length and hardness. Most war arrows were a full meter in length because it held a heavier arrowhead and balanced well for the length and combined thickness or gauge of the shaft itself. There was much that went into an arrow and Michea wanted to be an arrow maker in his olden years.

It was advised by his favorite arrow maker that the arrowhead he showed would not be for him. "Oh you are much too little of a man for this one," chortled the old man, "but thank you for showing me such a fine arrowhead. I have only heard of them. They are the way of the future I see!"

Michea's smile never waned as he exclaimed, "I will always consider this as my keepsake, for my Mother made it just for me!"

"Well she made it for you for the future," smiled the toothless old man. "You bring it back then."

That time would come.

Chapter 8:

South Florida 1528 AD

Michea becomes a man

Five years passed since discovering the powers of the Ponctoan Spear. Huega had since made several trips to the spot where the original obsidian rock was discovered. Many other chunks were dragged back to the settlement and literally thousands of arrowheads, spear-points and other assortments of superior tools made. Improvements in design could be seen in the exquisite clothing and jewelry as well as elegant carvings and many other untold items found throughout the Ponctoan society. Samples of obsidian arrow tips taken from their wounded and the dead puzzled the Calusa Indian. They were not without curiosity for the sources of this great stone.

One of the best of all improvements came from a small lad who hunted at the age of five; as skilful as those at the age of twelve. However, now that Michea was fifteen, he far surpassed those of over twice his age and experience. Today, Michea was being celebrated as a member of the tribe of warriors, and what an honor this was to both Michea and the elders of the tribe. The Chief was proud of Michea's heroism and his uncanny skill to outfox his enemy and ultimately save the village from a Calusa ambush.

Michea was humbled by this celebration and the honors bestowed on him, giving credit to the consultation of the Chief and the Ponctoan Spear. However, for Michea, his humbleness hid only his sorrow. A sorrow he brought down on himself. He felt as if his heart was being torn from his chest. He had brought evil into the village and there was a great nothingness in his soul because of it. He was now left a 'half person'; a person living out a life of 'no happiness'. On the bright side, it was a great victory. This was one of many, now that the elders learned how to manipulate the 'spirit world', by knowing in advance how events would play-out during the course of any one day in the future. This, Michea seen was part of the formula for his successful raid on the raiders who plotted to attack his people one night last week. The Chief warned of the ambush a few days before it happened. He recalled it with pinpoint accuracy and even told of whom headed up the attack. It was the Warrior Prince of the Elite to the Calusa Tribe. He was an outsider from a different land. A very clever and strong warrior, he was a cruel man who never took prisoners. He was one to strike hard and run when everyone was either unexpectedly away on hunting parties and the woman alone, or all asleep during the evening. This man was always accompanied by several of the best Calusa Warriors and he had consistently been unstoppable. He carried with him strange and powerful weapons.

On those days leading up to the attack, the Chief gathered the best of his warriors and walked them through the way the attack would come about. He would be certain his Warriors would be on hand well in time to ambush the ensnared. As Michea recalled, the Prince led the sneak attack and in the pattern, he saw, the only retreat for this foolish man was not on his flanks, but in the rears! And that is where Michea silently hid in the darkness, with only himself and the specially tooled obsidian sabre. Michea's sabre was a curved slice of carefully crafted obsidian, roughly three hands in length and bound to a finely turned cypress pole six feet in length. Michea found he could lop the antlers from an eight-point buck with a single swipe!

The battle ended nearly as quickly as it began, with the sound of arrows singing out in the night; when all throughout the sawgrass meadow that rimed the bulrush to the north side of the Ponctoan Settlement, silence fell suddenly. Michea smelled the pungent odor of smoldering wet grass or wood. Soon it would reveal it was the glowing red wick of a weapon that made a thunderous noise which both terrorized and killed. Quietly forward, in the moonless night Michea spotted the Prince! There he was backing up toward him. He peered into the night waiting for something; a sign to attack? *Ha*, thought Michea, *his men were already dead, all eight of them! And he, the Prince of the Elite Calusa will be nine!*

Michea could see him quite well now. He could see the Prince lining up his weapon for a shot, when Michea spoke out to him.

"Will you beg for your life?"

The Prince froze, lowered his rifle to his side and said, "And you take me as your prisoner?"

"I'll take you as a hapless fool," replied Michea. "For anyone knows not to attack a Ponctoan at night!"

The Prince was truly afraid for his life. He knew he was surrounded by warriors he could not readily see. "Spare my life and I will give you wealth and high standing in my Society!"

"Careful what you say," warned Michea. "In the past you have killed many of our innocents. Now drop your weapon and walk toward the village. Know I am right behind you ready to take you in dead if you desire!"

Something wasn't right. Something went wrong. Not with the plan for it worked well. It was apparent that the Ponctoan war party had departed for the village leaving Michea alone with this jackal of sorts. But then, *who was to know he was here with this jackal?* So it was now a matter of time before his enemy prisoner discovered this, and he did, as the Prince spoke first ...

"I believe your friends have abandoned you!" The Prince turned to face Michea for the first time and laughed. "Well, you aren't but a half warrior. I have weapons older than you!" And with that the Prince began to walk back to where he dropped his rifle.

Michea never said a word. He just flipped his war sabre upside down and silently swept it up to run a neatly applied incision between the Prince's legs near the crotch of his hind-end severing the carotid artery. Now Michea only stopped in his tracks and waited for this one to die.

The razor sharpness of the obsidian edge never registered pain to the Prince as he kept walking and searching in the sawgrass for his rifle. Ten seconds ticked by. The Prince picked up his rifle, turned to face Michea, took a deep breath, blinked his eyes a few times, and then leaned against his rifle as a crutch. Michea guessed it may take up to twenty seconds for this one to die.

Fifteen seconds.

"Guess you must be feeling weak," Michea said as he approached the Prince. "You are a strange man, but I must tell you, I have set you up to bleed-out." Michea reached for the leather thong around his waist. "And with this strap I will tie your legs like a game animal and drag you back to my village for display!"

The Prince was too weak to make a verbal response. He just crumpled to the ground to draw his last breath while Michea strapped his ankles together. "Hmm ... twenty seconds, maybe thirty." Michea took a deep breath and before dragging home the Prince he gave him a few last words of encouragement. "May you become a buzzard in your next life. There are plenty of your warriors out in the field for you to feast on. Eat hearty my friend." With that, Michea picked up the Prince's rifle and other belongings and began dragging his trophy to the village settlement.

Michea found dragging this Prince easier than he expected. It seems the torso armor helped lesson the drag through the sawgrass and even though parts of the swampy bulrush before reaching the settlement. He was greeted by his fellow Warriors as he broke through the thicket into the camp. Many questions were asked of Michea as he dropped the Prince about the center of the village circle.

"What was it like to make your first enemy kill," asked one Warrior.

To Michea, it was no different to him than killing his first bull hog. To kill a beast that kills just for the sport of it, like this Prince, made it of no consequence other than making his people safe. *It is what warriors do.* Nevertheless, Michea did not want to dwell on it. This man deserved to die and Michea was not remorseful whatsoever.

"What will you do with him," asked another.

"By first light I will drag him to the banks of the Peace River and let him float home to his people."

"Why is that Michea?" asked another voice.

"Should he make it back without being eaten by alligators then he will serve as a message, that attacking the Ponctoan People reaps consequences on the Calusa tribe."

In the village center, a hearth of coquina stone was lit in honor of Michea. For the next few hours the Warriors would gather and unwind from the battle by telling wild tales of other battles they had had.

Michea sat staring beyond the flames of the fire pit; seeing in the amber light the face of Chickpea.

She sat with other girls her own age. She was now nearly twelve years old. And every time Michea's eyes fell upon her face; her eyes would shyly dart away.

Michea only understood that Chickpea was quite reserved when being around him lately. Gone seemed the day when they played together. And gone were her silly face-making antics. She seemed so intellectual now. Many times he saw Chickpea doing significant adult chores, like tending to real babies and washing clothes. These things made her look mature. And in a way, Michea thought, very charming. So now Michea wondered if she noticed him as good-looking. He thought not. Girls have no such thoughts of men. He was certain …

"Michea," yelled a warrior from the circle around the fire. Michea snapped out of his daydreams of Chickpea. He raised his head to see a fellow Warrior holding the dead Prince's rifle. "How do these thunder weapons work?"

Michea wasn't an expert on such things as this. And as he watched his fellow warrior fiddle with the rifle, all the while it suddenly became apparent to him, what the results of such foreplay with this mystic stick of death would reap. Whom-shall-ever this evil thing may point to will die; and as quick in the drop of the trigger hammer, it came to pass.

A clap of thunder rang out and one Ponctoan Warrior was launched into a somersault as the rifle spit its revenge, into the circle of the fire pit and out of the warmth of the hearth spilled forth a cry.

Lingering in the gun smoke it became dreadfully apparent; the Prince had gotten his revenge on Michea. As swiftly as Michea had taken the Prince's life, the Prince had taken his. Michea fell back at the sight of Chickpea being blown-away by a single bullet. And he cried out to the spirit leader to take his life instead. No one noted his cry as all were shocked by what had just happened.

Michea held Chickpea in his arms and told her he loved her with all his heart. She in turn, while still conscious looked up to his face and smiled. "I have always loved you Michea," she said feebly. Michea sealed his wishes with a tender kiss, "You are my Lichea now," he told her.

In all the chaos that followed around Michea, little Lichea was cruelly plucked from his arms. It was her father, Huega. He cried out for the Doctor and also the Chief. Luega, Lichea's mother, collapsed at the sight of her little girl's lifeless body in the arms of her husband in the doorway to her family's hut.

Clugar, the hapless warrior who fired the fatal shot that killed Lichea, sat horrified by this event. He laid the weapon down and stood motionless for a moment before dashing off into the night wilderness. It was up to Michea to stop Clugar from exiling himself. There was no one else, as Lichea's father and mother had already begun grieving for their daughter. A life for a life was Ponctoan Law and even though this was an unfortunate accident, Clugar would present himself before the Chief for deliverance of a tribal law that would decide his fate. There could be no defense for such an act, even though it was an accident. Michea knew this. He also knew Clugar. He had grown up with him and knew he is a noble and trustworthy soul who did not deserve this curse that *descended* on the settlement. Clugar was merely as much a victim as was Chickpea.

Michea dashed through the crowd of villagers who formed the village circle hearth. He was bound to catch Clugar and when he did, he tackled Clugar to the ground. He held him there with all his might. He shouted in Clugar's ear that destiny is where spirituality lies. And that nothing that happened out of accidental events could reflect badly against him. Clugar flipped himself face up and tossed the lighter Michea *ass over a tea kettle* into the sawgrass a full two meters away.

He stood up and shouted at Michea, "You know nothing about spirituality!"

Michea charged out of the sawgrass to give Clugar what he needed right now and that was a swift kick to the belly. It was followed up by a few good slaps to the face. "Clugar," cried Michea, "I will not stand and watch my Warrior Brother be destroyed by the curse of an evil Prince. I will tell the Chief, I will stand up for you!"

Clugar was allowed to get up and face Michea. "You will die for those words!"

"If it is so … then it is so."

Clugar wouldn't answer, but instead charged at Michea, ready to pound him into a bloody pulp. Michea met the challenge in the flurry of fists by grabbing Clugar in a tight bear-hug. Clugar countered with a death grip that would surely snap Michea's neck and drop him dead as he stood now breathing. It was not only true, but very obvious Clugar's age and experience was far more advanced than that of one young warrior. Michea knew it. Clugar knew it. In that brief moment, confidence in friendship arrived.

"You swear it?" Clugar asked in an exhaled breath. It then dawned on Clugar that Michea was his forgiveness and confidant. He needed Michea to help him with the pain of what just happened. In the *warrior way*, Michea did precisely that.

"As I am your brother, I swear it so," Michea said firmly. "I hurt more than anyone you can imagine. And so, we shall grieve together."

Michea never having seen another Warrior shed a tear watched as Clugar wept openly and crumpled to the sawgrass. Michea knelt beside him and patted his shoulders. Strange, Michea thought. His heart hurt the worst yet he never let a tear. In the night someone behind him approached. It was another warrior who spoke boldly. "The young girl has been delivered to the Spirit World."

This was something new, to not only Michea but the entire tribe. Many of those who passed away would be delivered by the Doctor to the Spirit World. Given enough time, perhaps the whole tribe would disappear into the hereafter called forth by the Doctor and Chief. If they can visit there and return, then Michea thought he too would learn the ways of walking on the *spirit-side* someday.

With the celebration at hand the following day, Michea knew that this would be *his day*. Perhaps the only chance he would have to confront the Doctor and get answers. Now that his love was taken from him, maybe the Doctor can help him understand. Still, Michea was timid about his requests and kept a civil disposition. He gave a speech of encouragement to those coming up in the ranks as well as those like himself struggling to keep the settlement safe from danger. Strangely, the celebration went forward without burden of the memory of the former Chickpea's death. And it began to wear on Michea's faith in all things true in life. It was not right that he should celebrate life and partake in the feast without her happy face in the crowed.

Especially when her last words to him were, "I love you!" This alone meant something. This alone meant she wanted to spend the rest of her time in this world with him as his wife. She would have made for him a family, a son perhaps. But now she is gone. His Mother Waukee had said it would take time for such wounds to heal. Michea didn't want time to heal this wound. He needed the Doctor to help him make things right!

Lately, it has been noticed by many in the tribe, the Doctor aging rapidly. His hair getting grey and heavy lines wrinkled about his mouth and eyes. It seemed the more he visited and the longer he stayed in the *spirit-side*, the older he became. There was whispering in the tribe that noticed the Doctor becoming old before his time. Michea wondered was it only the time he spent in the spirit side that made him older faster. *There had to be a difference in the scale of time between both now and the spiritual world*, Michea thought. *Would the Doctor talk of this to him*, he wondered?

"Where is Lichea," demanded Michea of the Doctor.

"You mean Chickpea?" the Doctor responding with raised eyebrows.

Michea eyes softened. "Before she was taken," he said solemnly, "we sealed vows."

"Is that so ... and where was the Chief to consummate this union?"

"There was no time," Michea snapped.

"Then it was consummated under the eyes of God?"

"It was," Michea said without hesitation.

Now that marriage entered the picture, the Doctor knew Michea was right. It was a timely exchanging of vows between man and woman. This, he knew, must be respected. However, there were the formalities that had to be followed. Michea would still be required to ask for the hand of Chickpea and in a post-posthumous ceremony to follow, a proxy of Chickpea would be appointed to *stand-in* for her.

"There are many things to declare in your life young Michea," the Doctor said. "Before you can be taken seriously, you must follow-up on your responsibilities and accept your burden in life."

The Doctor studied the angry eyes of Michea. He drew a weary breath and looked aside for a moment. "Don't burden yourself with matters of which you cannot understand."

"I must know," Michea demanded. "It is my day and you owe me! Tell me Lichea is well and happy in the spirit world ... tell me she is not suffering ..." Michea was surprised to see the reaction of the Doctor. This was bold language for any man to put forward to a holy man of wisdom and infinite knowledge.

"Enjoy your day Michea," said the Doctor. "It may very well be your last."

Michea's heart jolted to a halt it seemed. The words of the Doctor he had never heard before but only imagined. *And for him to say this ... a warning?*

"Go and enjoy what of it you can Michea. Tomorrow there will be answers I promise you," the Doctor said softly. "My journeys have come to pass and I have known my journeys."

For most it has been said, to know the past ensures one not to repeat it in the future, but for the one who knew the future would surely not repeat the past. Michea wondered if it would not be wise to try to convince the Doctor that he needed an apprentice. If it has been seen that bad events can still occur, with one observer, less evil to happen with two observers! So in the following day, Michea rushed to the Doctors hut to describe his ideas. The Doctor listened courteously and in turn graciously declined Michea's proposal to be his apprentice.

"You are too young Michea," the Doctor said. "But I will meet with the Chief later and discuss your proposal."

Michea nodded his head. He was happy to have this discussion with the Doctor. And now it was going to be discussed with the Chief? Michea was feeling special and harbored no resentment for being told he was too young. Especially after seeing how rapidly the Doctor aged. Given the time, Michea would soon come of age on his return from a few trips to the *spirit-side*. He wondered if it hurt to age this fast and since he had the Doctor talking freely to him, he decided to throw that question out there.

He had at no time seen the Doctor laugh with such merriment. "You know Michea; I have never given that one a thought!"

"I don't understand." Michea stood before the Doctor. He couldn't help the puzzled expression on his face. And thought for a moment the Doctor looked puzzled as well!

"Well, you know," the Doctor began, "I never thought of it like that. That is the why our people see my aging quickly. After all, I only appear to be away to the *spirit-side* for a few days to maybe a few weeks at a time. So they must, like yourself see me aging very fast when in reality, I have known it to be years!"

Michea was astounded. He was speechless for a moment. "You spent years in the *spirit-side*?"

"Yes," he replied with a sigh. His eyes sparkled with a slight grin. As if he was recalling memories of the years spent on the other side, "and I have learned more than a hundred lives!"

Michea eyes lit bright. "Oh Great One, tell me of the lessons you have learned!"

"I shall," he said earnestly. "But you must do your chores and learn more of your time right here."

Michea nodded without a smile, and then walked proudly out the door of the hut. The Doctor on the other hand, pondered his two lives. He realized suddenly that he had spent more of his life in the future then here in the *now*. In all his time travel, he learned much. Nevertheless, there was still so much of the unknown to discover.

In his longest of all journeys, he learned of *what must be and what was*. For instance, he learned of the Spear and how it was formed. He took to the sciences and the libraries, learning to read mostly on his own. Of great cities, he dropped into learning centres called Universities and argued with scholars on topics related to Relativity and the works of Einstein, and Galileo, and Newton. Some made sense to him.

Even so, nothing ever explained what the wise men of the Universities rebuffed on the optics of one to see portals in time. Nothing to explain the bubble trapped in obsidian. The Doctor never allowed anyone the sight of the Ponctoan Spear. It was his only way back home he knew. And he kept it on him at all times.

One Professor pondered the arguments and theory the Ponctoan Doctor presented; in that of all Black Hole properties considered in an abstract math model. However, something was missing. Not that a Black Hole had to be of any one certain size. Not at all and that they could only be found in outer space. Contrarily, they could possibly be very tiny and floating about in our atmosphere! So small to be seen or contrast with the naked eye. They could be broken fragments of huge systems that spewed tiny holes like eddy-currents out of a watery whirlpool. They would almost have to be attracted to wander about the surface of the earth. But what would that attraction be?

The tribal Doctor sat following along with the conception of what black holes were and how they attracted and sucked in light around them. And even for their size they had remarkable power over the fabric of space and time.

Then one day the Doctor recalled the straight line that the black holes followed throughout the village. There was a few other lines' way off, in the distance. However, they could only be seen once one was on the other side in the continuum of times' stone walkway. Stone walkways that crumbled away the past still puzzled the Doctor. He could not crossover into yesterday. And he could not cross-over to the other walkways, because they did not seem to cross-over one another. Perhaps somewhere they came close and one could leap over to the other from some point. But who knows what would happen if one fell in the attempt? The events from that far away walkway could not be seen with any certainty to know the answer.

To the professor it was an interesting point. Of what formed these long walk ways. And if they existed it would have to be a complement to the properties we already know of black holes. Magnetism perhaps?

The professor gave a hoot and clapped his hands with a revelation that proved the tribal doctors theory. It was in the theory that the earth had many magnetic fault-lines. The question of where these fault-lines were remained a mystery.

This marked the very first month of a twenty-year stay in Cleveland Ohio for the Ponctoan Doctor. It also marked nearly a hundred yards of a walkway in the time-continuum to crumble into infinity. The Doctor saw this as simply a loss of two months of his tribal past. When he did return, his people had only seen him gone a few months.

Even so, for in this moment of discovery, he and the professor decided to move the discovery to a local pizza shop right next door to an all-night laundry-mat where the Doctor had spent the first weeks of his stay. *That was nearly two years ago*, he thought. He had seen and learned so much!

Drinks arrived at the table first, a couple of colas and a few tumblers of ice water. The Doctor felt at ease with it all, especially the cola. He had sampled his first one left behind by a lady in a bus stop. He saw her enjoy it and it looked very good to him. She must have noticed him watching longingly at the cola and so she asked him if he wanted it since she could not take it on the bus. So his initial encounter with carbonation was to naturally blow the first swallow out his nose. Every day thereafter he made sure to be at the bus stop at 2 pm weekdays. She'd leave him the rest of her Coke before hopping the bus on her way to the hospital where she worked.

The professor excused himself for the bathroom. The Doctor thoughts wandered off in the thought of becoming a city dweller himself.

The Doctor spent his first few weeks as a new city dweller just sitting and observing from various bus shelters on the city's west side. And of course when the weather got rough he found refuge in the city's bus garage. There he found an entire colony of buses. Most parked inside an open garage structure. It was huge and there were buses as far as one could see. The seats were soft and inviting. The Doctor had never seen such comfort!

Soon the opportunity to make money by becoming a custodian presented itself. It came his way via an old entrepreneur who owned a cleaning service that he ran from his van. It was evening work that he did three nights a week. Fizzy (the owner of the cleaning outfit) and two other young fellows, Paul and Bobbie traveled throughout the city, going from one bus garage to another cleaning the mechanic's bay floors. The routine was a simple one. Go in and first move the tools and equipment aside so that the floor could be hosed-down and soaped. Then, Bobby Paul and the Doctor would a scrub the soapy mixture around with a broom until the grease and dirt was loose enough to hose it out to the street. After that, the remaining water was squeegeed away and the mechanical equipment put back in place.

"Easy money, eh Doc," said Paul.

The Doctor looked at the money in his pay envelope, all cash. He looked at every bill. It was the same and there were five of them. Each with the portrait of a man the Doctor would one-day read about, George Washington. It was very good money for the time-continuum he assumed. It was easy money and with it, he could now trade and barter.

Beer he found was not worth his take-home pay for the evening, so he waited until the next morning for the purchase of a tin top he saw in a toy store not long ago. He saw a young boy spin the top by drawing out a metal rod from its center and pushing it back into itself with a rapid up/down motion. The top spun in place making a beautiful hum. Mesmerized by the action of the top, the Doctor studied it and all things about it. In many ways, it behaved like the black hole theory.

Even so, there were other fascinating things like the painted people with the wild red and purple hair and big bulbous-shaped red noses! There where animals and bright colors painted on the top as well.

Later, during a weekend, the Doctor found a place called the Cleveland Zoo. The world was becoming a big place; much bigger than the Doctor ever imagined. To have wild habitats to support such a variety of game was mind boggling! There were apes and elephants originally from a place called Africa, tigers from India, and mountain goats from Switzerland.

It was a wonderful day of discovery. It was the day the Doctor discovered his home in South Florida, to where he stood, was a world apart. He began to wonder what it was like there now. It would take an airplane or a bus to get there and that he knew. What he didn't know was how to go about organizing such a trip. If buses weren't free, then it was certain that neither was airplanes. Maybe he would ask Mrs Barbara, who lived up the stairs from the coin laundry-mat. She was a kindly middle-aged widow who loved to talk. She was fond of the Doctor and had given him a break on the apartment rental. He moved right in without a security and last month's rent deposit. *She was indeed a sweet lady*, the Doctor thought. She had plenty of lady friends in and around town and many from church who sent nice used clothing just the right size for the Doctor. And since the Doctor had once started young, as an apprentice arrow maker, he was good with his hands. He had an ability to study something and figure out how to fix it. Such was the case with those little things that would often break around the house. Loose or broken screens, creaky doorsteps et cetra.

He also learned that people were people no matter where you go. Even through time it held true. Of course true human spirit really comes alive when survival becomes easier. Then there is plenty of time to get to know one another and become peaceful and caring.

A warm feeling of security flooded the Doctors senses. There was hope in this future. The professor had just returned to the table where the Doctor was still engaged in retrospection.

Now the Doctor had the opportunity to sample all kinds of futuristic foods. Pizza was not one of them. It looked good but it was too bloody, so he sent a few slices back to be cooked longer. The professor was a portly man who gobbled up the remaining pizza. Then finally after a few send-backs for more cooking, the slices were black.

"Man," the professor said wiping his mouth on a napkin. "I don't know how you can eat burnt pizza!"

"It is not good to eat so much blood!"

"Blood?"

"Way too much blood," the Doctor began. "Look, it's even on your blouse ... that's too much blood!"

The professor muttered expletives and grabbed a napkin and dipped it in a water glass. As he was trying to clean his shirt the Doctor asked, "Why do you curse God?" Ciabee did not like this man. No good man makes evil the name of his God.

"It's a 25 dollar Arrow shirt"

"Funny. Not look like arrow."

"And what's with you and tomato sauce? I thought you were Italian?"

"You say I was Italian. I just say Okay. What's Italian? Same as Indian, right?"

The professor rolled his eyes back. "I got it," he said. "Hollywood hires Italians to play the roles of Indians. In real life, it's the other way around!" The Doctor just stared blankly. He hadn't a clue. "It's a joke," he sighed. "And oh, the pizza wasn't bloody. It was ... my god; don't tell me you've never had a tomato!"

The Doctor thought a second. "Yes I have heard that word before. They are good. Small red circles with seeds. All melt in mouth."

The professor threw his hands up and said, "Okay let's stop farting around here. You in an odd sort of way are very up on this subject of black holes.

And you think I am on to something when I suggest magnetic fault-lines? I mean after all, these small black holes would be very susceptible to gather near these attractive forces. Nevertheless, I've got to tell ya doctor we've got to show proof. I believe I have enough to get a paper published on this ... but at this stage, it is merely a concept."

The tribal Doctor was still learning the language; including the language of slang. He hoped he was getting across his point accurately. "If you develop a way to see these fault-ways, would that be proof?"

"Something like a divining rod you said?"

The professor could see the Doctor was getting confused. "What kind of Indian are you, by the way?"

"I am Ponctoan, by the way of South Florida." It was beginning to wear on him, this learned man of manipulation. The Doctor was becoming suspiciously cautious about this man. *Was it not the black doorway of time? Knowledge the people of this time already have? Is it I, the only one, to be learning of this knowledge,* he wondered. It was this revelation, the Doctor discovered; *he'd had all the answers after all!*

And then there was something in the way this man carried himself, the click of his pen and the rapid scribbling of notes that made him feel uneasy.

At his side, he felt the presence of the Spear. It was there in his pocket where it had always been. He never went anywhere without it and when at sleep, he kept it safely tucked away. "Divining rod," the Doctor muttered, still in cautious thought for what he would say next. "And this Divining rod, what does this rod do?"

Paging back through his notepad, he answered numbly in other thoughts, "It's a stick used by plumbers and well drillers to find water ..." The professor sighed a moment, still lost in thought.

It didn't make sense to the Doctor of what the professor was leading up to. Nevertheless, a stick to discover water had nothing to do with finding these so-called black holes. Non-the-less, the next question the professor asked, did make sense and it hit home to his tribal past.

"So you said an instrument of which you can peer through can see these tiny black holes?"

"I said not of this."

"Well, you said a stone spear-point had such a lens upon which you could see small portals of light, or rather points of darkness?"

The Doctor could see the writings of which were loosely his and the observations of the professor. It was all written down. *He spoke of this a long time ago and thought that might have been a mistake.* For now he could see the primeval ambitions that had stirred in this one, ambition for power. And there, in the Spear was plenty of potential for it. "There is no time travel ..." the Doctor said stiffly, trying not to make it sound like a question. The professor stopped peering thoughtfully at his notes and looked over to the Doctor.

"I didn't mean to imply that. But it could mean that the possibility does exist. You do understand as a doctor that possibility exists, right?"

It was the answer that the Doctor needed. It was true. Time travel does not exist in this new world. And now, it was the conviction that time travel should not happen in this time. It was a thought brought up from his heart, that it should *never be.*

"So tell me where you learn of this Spear?" asked the professor.

"It was not," the Doctor replied evenly. "but a thought. Or theories as you have said before."

The professor shook his head and blinked his eyes in disbelief. "Wait-wait-wait, a minute," he exclaimed.

The two of them sat there staring at one another. Again, the ambition showed in the eyes of the man across the table from him. Some things never change. Even after half a dozen centuries, man will always be man in many ways, including evil.

The Doctor leapt forward and slapped his hand over the professor's notepad. He then bit his lip and slid his hand away, sinking back to his seat in the booth where they sat. "I don't understand why you are getting upset with me," the professor said. "If it is because you are preparing or proposing a scientific manuscript on your own … so be it! Just say so. I am not into plagiarism. I am a respected scientist and professor and would like to keep it that way!"

The professor was getting upset with the Doctor and was reaching for his coat by the door. "I'll tell you one thing," he advised the Doctor over his shoulder, "you're not going to get anywhere without proof of theory!" He then turned about to face the Doctor. Pushing a finger in his face, he arrogantly advised the Doctor, "And you aren't going to publish this without me, Professor Maxwell Tanker, to co-op your manuscript!"

There would be a time to argue, but not with scornful ones like this man, thought the Doctor. And yes he was an arrogant soul and no doubt foul-hearted. It was likely this man to pursue the Spear putting the Doctor on the game trail out of his greed to obtain personal wealth. The Doctor felt as if he were the game and the professor the hunter. *Two years here was long enough,* he thought. It was time to go.

For Miss Barbara, there were just few things she hoped Ciabee would tend to. Yes, the Doctor grew tired of Miss Barbara calling him Doctor since he grew fond of her hospitality and her baked chicken. Her pork chops and scalloped potatoes weren't bad either! So before he left, he would pick up a few night classes at the trade school across town on Loraine Avenue. Soon he was fixing washing machines and dryers. That really helped a lot and kept Ciabee busy in Miss Barbara's downstairs Laundry-mat. Now and then, the time came for heading back to the tribe, but then Ciabee looked down to the gold band around his finger and the beautiful face of his wife Barbara and smiled knowing he was home after all.

Barbara was unique a lady to Ciabee. She had an undying trust in him and didn't care that he came from another time. Whether she believed it or not, never seemed to matter to her. Her trust in him made him love her even more as time went on. Though they never had children, it at no time mattered to Ciabee. He was getting too old for fatherhood by now and the upstairs apartment needed to be refurbished anyway. *Maybe he would knock-down a few walls and make it all one big house for the two of them!* While doing so, Barbara discovered the toy tin top; Ciabee had bought several years ago.

"How cute," Barbara said looking toward Ciabee, who was busy painting baseboards. "It is yours?"

Ciabee looked up giving it a passing glance and then back to painting. "Yes," is all he said.

"How long have you had it?"

"About the time we met."

"So you bought this as a gift?"

Ciabee stopped with the painting. "No not really, I thought it was interesting so I bought it."

"Oh," she said, looking as if she was certain Ciabee was hiding his intentions for the toy. "I thought maybe you were anticipating a child."

Ciabee knew. *You only have to live with a woman a decade or so before you can read her mind.* Even so, Barbara was his soul mate. It was destiny that joined them and he knew she wanted to tell him it was her. She was the one who could not conceive a child.

Barbara's eyes shifted to the ceiling. It was hard for her to put this together, but she figured she would have to tell him someday. This was the day. "Ciabee, you never really told me you wanted a family," she began while trying to keep her tears from spilling down her cheeks. "You always said it didn't matter either way to you, but this toy and all ... I just think I should tell you that it is me. I can't have a baby ..."

Ciabee took her by the wrist and pulled her down next to him. He held her and told her that she was his baby. "Believe what I say," he said firmly. "The toy was for me!"

In a brief moment of silence, they both burst into laughter and suddenly, nothing mattered, no babies nor toys. No one loved each other like Ciabee and Barbara. And since 1952, their love endured for over thirteen years.

Barbara fell ill one Saturday. Ciabee became worried about the way she coughed. He had an ambulance quickly transport her to the nearest hospital just a few blocks down the street. By Monday Barbara passed away. Ciabee was devastated. After weeks of grieving, Ciabee knew his heart would never heal and his memories of Barbara would always be with him. There was no sense torturing himself by remaining in their home above the laundry-mat. So with much regret he left it all behind.

It was now time to see if he could find his way back to his tribe. It was so long ago, Ciabee had thoughts that things may have changed and his re-entry into the time-continuum may have shifted or worse, closed up. Who knows?

The magnetic fault-line that ran through a cemetery and across Pearl Road into a neighborhood was the place where he recalled having entered his time here. Ciabee decided to disappear into the time continuum in the cloak of the night. Fewer chances he would be seen.

He had fielded calls from Professor Maxwell Tanker over the years. He was a persistent man indeed! So Ciabee would not want this man to see him with the Spear on his person. It was a big city Cleveland; but too many times, Ciabee had seen Tanker … and Tanker was watching. Although Tanker's messages were basic and non-threatening, he still wanted to get together with Ciabee and discuss black holes. The competition on black hole theory was a fierce one that had been whispered about for years but never proven.

Tanker needed something to prove his theory. And Ciabee had that something. If there was a magnetic fault-line close by where Ciabee lived, Tanker had dedicated his research to having Ciabee lead him to it. And since Tanker knew of Ciabee's wife passing away, it would make sense to him that if he did have knowledge of such a fault-line, he would use it to slip back to his tribe.

Tanker was a diligent researcher in every angle possible. He researched the Ponctoan Tribe and found very little about it. However, what he did learn made him very suspicious about a man claiming to be a descendent of such people who had literally vanished without a trace! Unlike the Mayan people or Incas, the Ponctoan people disappeared leaving behind a number of witnesses. And Tanker could uncover some pretty good evidence that these people were time travelers! He just recently pieced together startling evidence in South Florida of what could only explain these powers this Indian talked of. A magnetic fault line and this very fault-line may have run directly across the tribal settlement of the Ponctoan Tribe. Not all fault-lines travel in straight lines, while most move slowly over many centuries. So it could be assumed, if a location to which a known fault-line exist, this line could be plotted along magnetic degrees of a compass. Still just another theory, but Tanker had done his homework and his assessment showed roughly the area Ciabee would eventually show up and make his departure.

Tonight was the night and the symbol to tip off Tanker, was Ciabee had left his flat dressed in his traditional garb. Going back the same way he arrived. Ciabee's buckskin vest, a fit a bit more tightly as it were a testimony of Barbara's good cooking.

Overhead, through the barren boughs of maple trees, lining the cemetery row, held captive a moon, the size of a silver dollar. A ghostly wind howled and swirled the leaves about in tight columns that pitched up like dust-devils, then dissipated off here and there. With each gust, came a misting rush of tightness across a bare chested Ponctoan Medicine Warrior. Ciabee shook-off the cold blast, keeping his concentration on a place where he came to know as his entry-point. It was in the cemetery where Ciabee recalled an angel appeared. A beautifully sculpted angel in white stone, suddenly he felt a warmth flood over him; the feeling that he had come back to the right place. Turning his head about as if to bask in this warmth, he thought he spied a dark figure way off to the street-side area of the cemetery. That one, he knew was Professor Maxwell Tanker. He stared for a moment never realizing whether or not the man was approaching him. Only that, he needed to leave right away. He raised the Ponctoan Spear-point to his right eye and peered directly through the bubble remnant of Trilo and the miracle of the way the obsidian had been cleaved into an optical lens sufficient for the job. And truly the job at hand was seen. A stream of tiny holes bounced and floated close by. Sparkly little circles of jewels surrounded Ciabee. He felt fortunate. Even lucky as he slowly looked away from the Spear-point and whispered Cleveland good-bye.

"Farewell my darling wife Barbara," he said softly.

With that, Ciabee knew a chapter in his life had closed. Moments later he was standing on the stone walkway of the time-continuum that was indeed the walkway of this magnetic fault-line. The one Tanker had talked about.

Tanker got close enough to see the whole thing. The Spear-point, the brief flash and rip in the fabric of space and time as Ciabee ripped open the hole and stepped inside. Maxwell Tanker was taken aback at the sheer power of these small black holes and yet, the only way to manipulate them lay in the hands of an Indian; an ancient Ponctoan Indian.

Maxwell Tanker would never be the same man. He would be an obsessed man searching for a way into the time continuum, though first, he must discover the secret of this Spear.

Chapter 9:

Davistown Hospital Pennsylvania.

Michea's bid for survival

"Sir," a young female voice echoed off in the grogginess of Alan Wheeler's head, "sir?"

Alan barely opened an eye. He had been asleep a long time he knew. His left arm bandaged or in a cast, he slowly tried to gather what had happened to him. And as he began to yawn, he felt a strap of gauze-like material under his chin. The material ran up to his forehead where it circled his entire head. That's right he said thoughtful, "I must have banged my head."

"Oh yes you did," said the female nurse, "and dislocated your shoulder too!" She moved around to the other side of his hospital bed to gather a blood pressure reading. "You tumbled out of an ambulance last night I hear."

"Oh my god," Alan exclaimed, "the man that was transported here!"

She cut in quickly, "He's going to be fine," Raising a finger to her lips she focused on her BP reading.

"Fine?" Alan quite puzzled, raised his voice. "He went into cardiac arrest!"

"Calm down," she said. "He is resting well as we speak. We have him in Intensive Care right now, but I understand he will come out of it. He is remarkably strong."

Alan was informed, that only after he had been discharged from the hospital, would he be allowed to visit Michea. But in the meantime, he would be asked a slew of questions about this young man. Alan also had his own questions about his friend Michea. "Okay," Doctor Williams said. "Your questions first."

"Why did Michea go into cardiac arrest?"

"A combination of complications, such as extremely high temperature, blood pressure not to mention severe dehydration."

Alan waited for more. Doctor Williams shook his head. "He apparently has a severe reaction to the flu. Zero antibodies. He is like, I recall, from early studies of Indians, very susceptible to even the common colds. I know that sounds crazy, but that is what the lab reports show. But there was something else about him that troubles me ..."

"You know I am a friend of this young man, but I have only known him a couple of days." Alan said. "So I don't know much about him."

"I am supposed to involve the police when I suspect someone has been involved in a crime."

Alan shook his head clueless of the matter. "What do you mean Doctor?"

"It appears from the x-rays of this man, that he had been shot. And it really isn't obvious without the x-ray so I didn't notify the police."

"Good grief!" exclaimed Alan.

"I mean it's an old wound. But peculiar in that it was an arrow wound. He could have only been several years old when it happened. He's too weak to have the arrow tip removed."

Now Alan was getting more and more confused. "You mean Michea has a piece of an arrow sticking out of his back?"

"Not quite," mused the physician. "The arrow was removed years ago when it happened, but, like I said, it left the arrow tip in him. It's lodged above his shoulder blade and just under his collarbone. But I don't see it as a problem unless it begins to give him discomfort. In that case surgery to remove it would be ambulatory."

The doctor was about to shake Alan's hand and depart from company when suddenly he moved his hand to his forehead in a last minute gesture of thought. "And oh yeah," he said. "He also has a nasty looking knife wound in his side. I mean it's a scar, but quite a recent wound I would say!" Alan thanked the doctor and after shaking hands, the doctor turned and went about his visits.

This was all too strange. News like this led Alan to believe Michea had lots to tell him and not just about black holes. Since he'd been released from the hospital, Alan asked when he could come up and visit Michea. He was told since Michea was in a sense indigent with no indication of having a next of kin, Alan would be treated as Michea's closest relative. In other words, he was allowed to visit as soon as the Intensive Care physician seen fit. That could be in as little as an hour, so Alan went to the Intensive Care waiting room to await permission to see Michea.

Why not? After all, if anything Michea was alone in this world. Alan could associate with that. He was raised by his aunt and later at the age of five Saint Mary's when his aunt no longer could afford his medical bills. He was a charity kid and that was enough to bring down the scorn and the ridicule he suffered throughout grade school. He got better of course and began recovering from polio with the selfless help of the sisters at the school and orphanage where he stayed. No one wants to adopt a defective kid, but at the orphanage, he was whole and vibrant. The nuns of faith looked beyond his braces and made him scrub floors and wash windows in the convent. He scrubbed brass candle fixtures and took pride as they shined like gold! He had a purpose and later when his grades improved the nuns of the convent rallied for his graduation into higher levels of education. As the faculty of his advisers, they found the funding and the scholarships, getting him where he is today!

Yes, it was time for Alan to be *pay-it-forward!* Michea needed him and he was not about to let him down. He would get ready a space for him to live and recuperate.

"Mr Wheeler," a lady in the room called, "Mr Wheeler?"

Jarred from his thoughts, Alan perked up, "Yes?"

"I will take you back to see Michea now." She smiled and motioned him through a door that had an intercom wired to it. Pushing the button she asked someone name Aggie to open the door for her. Through the door was a circular corridor that circled back through another set of double doors and then to the Intensive Care Ward.

"Oh," said Clara the waiting room lady, "You are going to have to put on this."

She handed him a gown and gloves, "and oh yes this too!" Alan looked at this, curious object that resembled a shower cap. And lastly, a surgical paper mask. It was slow going for Alan. He was in pain and sporting a splinted left arm and a sling.

Clara saw this and immediately offered to help Alan. He graciously accepted her offer. It was her idea to safety pin the gown around his sling so that he could suit up easily. After all, Alan would not be able to use that arm anyway. So it made no sense to work his arm through the sleeve of a gown and then back into a sling. "Well there you go Mr Wheeler," Clara said jovially. "Oh and I think the young man is waking up!"

Alan's mind wandered away from Clara's comments. Alan noticed this room was posted as being quarantined. Then it dawned on Alan that it wasn't a warning for those who entered, but rather as protection to Michea from those entering his room. His immune system must be very fragile. Alan immediately noticed all the IV bags hanging about Michea's bed and all the tubes and monitors attached to him.

Michea stared openly at the doorway as Alan came into the room. Michea pushed himself forward from his reclining position to where he sat straight up in bed. "Doctor," he said weakly. "You must promise me," he began before being chastised by Alan.

"Michea for god's sake," Alan said tersely. "I am here. Everything is alright! Lay down and rest."

Michea would not stop. He was animated with a burst of energy that Alan knew would only last a few minutes. After that, who knows what may happen. "Slowly Michea, slowly. I am here and I am not going anywhere. Trust me kid."

He thought he saw Michea actually smile. Just as ordered, he lay back down, closed his eyes and drifted off to sleep. "Besides, it's a real pain in the butt to suit up and see you. So I will hang here and save myself the trouble."

Hospital PA calls echoed down the hallway. Soon it all faded away as Professor Wheeler drifted off to sleep

Chapter 10:

South Florida 1533

Michea's next big sorrow

Michea held favor with the Tribal Medicine Doctor. He visited often and brought the Doctor some of his favorite meals that his mother would help prepare for him. Even so, the Doctor sometimes talked about *baked chicken and pork chops?* And of a fine woman, he married a long time ago. A distant city in a valley, he'd say, far far away. Could it be Michea wondered, that he could remember the Doctor as a younger man? He may have been a man not much older than the tribes' eldest warrior; similar to the age of Clugar? Maybe as it was, the Doctor had seen too much sorrow in his time. After all, he was the uncle of Chickpea, now known as Lichea. It was he who had the heart wrenching burden of laying her to rest on the *spirit-side*. It was five years, yet he was still with Michea, mourning her loss. He suffered hard to speak her name, yet he had to know more about her life in the hereafter.

But there would be no talk of that, at least not now. There were many other things Ciabee wanted to talk about. He wanted to congratulate Michea on his victories while he was away. It was told that Michea had tracked a warring party looking to exact revenge for the death of their Elite Hispanic Warrior Prince. They came to pounce upon the settlement in another surprise attack. However, it was rumored, that there were double the number of warriors this time. The counter attack was coordinated with Clugar as seven of his best Ponctoan Warriors, Warriors in their prime. Michea would be the initial defense lookout and with luck he would have them ready for the traps he had set for them. The first being a deadfall trap rigged to drop four large coquina boulders on the heads of those passing through. The panther pit was Michea's most ominous trap. Set up to drop as many as four Warriors to their deaths, impaled on sharp bamboo stakes.

Michea scaled his way up and atop a tall cypress tree. From here he could see for miles all around him. He could see the Peace River make its bend around the first bay head of Mangroves before heading out toward the sea and as anticipated, the march of Calusa Warriors making their way toward the settlement. Michea counted eighteen of them. Maybe nineteen. They were still a day away, from the Ponctoan Settlement but Michea figured that maybe it would be nice to drop a few of their numbers from the list. Michea patted his pocket. In there was his lucky arrowhead that his Mother made him out of the first obsidian rock. Now ever so slowly and calculating, he drew an arrow from his quiver and nocked the feather end into the gut string of his ash wood bow. Then drawing a deep breath, he lowered his arrow to take aim. At this height in the trees, the arrow distance was shortened and the striking power of the arrow amplified many times over. The leader of the attacking party had just parted a cluster of broad leaf banana, when he was met by the zing of Michea's arrow. In a flurry of activity, Michea quickly nocked another arrow and let it fly. The initial arrow split the breastbone of the first warrior and thus he was dead before he hit the ground. The second warrior shocked by the attack of the first, went down with the force of the second arrow passing completely through his abdomen and into the shin of the third warrior behind him. Michea knew by now his luck was wearing thin. Soon he'd be the target of a dozen warrior arrows. Coming down from the cypress tree in haste usually created skinned elbows and cuts and bruises. He literally fell his way down from branch to branch. It never really mattered as to his covertness now, because the game had shifted to a new strategy. Michea literally landed straight into the path of his enemy! And as he did, he dove forward over the Panther Pit taking on a somersault that barely got him through without dangerously falling into the pit itself. While getting to his feet, he saw the first four warriors drop out of site into the realms of the pit. Only a dozen left he calculated.

He never slowed down but kept his pace far enough ahead to disappear from the sight of his enemy captors. A signal overhead in the fashion of a large clump of Spanish moss hung precisely over the path where his deadfall trap was rigged. Not far to the other side were his fellow warriors. In all there were seven of the best warriors in the settlement. All headed by Clugar, the tribes new Warrior Prince. They waited silently on guard. The time was about to come and all were arrow ready and waiting. Those remaining enemies who had escaped the pit worked their way around to re-join into a position not far from the dead-fall trap. One enemy warrior appeared to make signals to the others to take leave of the trail when Michea jumped up and let out a yelp! The only weapon he had readied was his obsidian knife. It looked as if Michea was challenging them all to hand to hand battle! Out of nowhere, the camouflage face of Michea appeared from a blind of sawgrass. The chase, commenced again. Michea dove off into the thick brush and as he did, he swiped the trigger to the deadfall trap. A loud snap and the ground thundered under him. Two more Calusa warriors met their end under an assortment of large coquina rocks. Just as the horror began to sink in, the remaining Calusa warriors met up with Clugar. His magical Ponctoan Warriors appeared out of the tall sawgrass, their faces striped cleverly to match their surroundings. In an instant as it would take to sweep the hair from your eyes, the remaining seven Calusa warriors were dropped like bull hogs in a blind.

The Ponctoan Warriors slowly emerged from their hiding places in the tall sawgrass. Clugar motioned to hang still and wait for noises in the forest ridge leading back to the Calusa territories. All at once, a flock of roosting birds took flight from a large Magnolia. "They got away," Clugar said his voice just above a whisper, "two, and maybe three!"

When all was settled, Clugar ordered his warriors to fill in the Panther Pit and disperse any evidence of the deadfall trap. The two who perished in the deadfall, were dragged and dropped into the Panther Pit with all the other dead Calusa Warriors. After all was done, the men sat around a trail fire enjoying a meal of otter, gopher, and steamed cabbage root cut from the heart of small scrub palmetto plants. When the meal was over, the men talked about many things. Conversation of the best way to put out fires that went out of control to silly subjects like how to get a duck to chase a bathing girl from a watering hole.

Michea on the other hand remained silent. He sat poking the fire with a long branch of red oak. "Michea," Clugar asked, "why you so quiet?"

Michea looked up from his thoughts and announced, "There will be another attack soon."

Clugar pursed his lips. He didn't want to hear of it so soon, but asked, "When?"

Michea slowly responded, "The night before the second moon's end."

Everyone in the circle raised their heads to look at the moon. Clugar thought a moment, "Are you sure Michea? That will be in thirty nine days!"

"Or by the end of the tenth day and one phase," Michea responded, by adding out the adjustment of tonight's moon.

Clugar knew Michea's words came straight from the mouths of the Chief and Medicine Doctor. They were the spiritual scouts that gathered this information. And without the aid of the spear point, survival in this settlement would have been impossible. The Calusa were getting impatient with the Ponctoan society. They wanted them out of this prime hunting land. The Ponctoans would not be pushed to the sea where there would only be ghost meat to survive on. To be taken captive and sold into slavery was the only option besides death. And only the young and strong would be taken, the rest murdered. It would be a two months journey on foot to Charles Town (known today as Charleston South Carolina) where slaves would be sold, so only the strong would survive that walk into slavery. Clugar shuttered at the thought as it raised the hair on his neck.

"How many Calusa will join in this raid?"

Michea shrugged his shoulders while still looking down at the fire. "Same," he replied. "Maybe more."

Clugar thoughtful of Michea's reply, sat back down to the fire. Now he too stared into the fire. The thought of it was ironic in that he and his men had already slain at least three times over the Calusa warrior numbers, than the Ponctoan had in total! Their Chief will become wise to this and will soon organize an all-out assault! Clugar's heart pounded. How many warriors would that be? Many more than Clugar could count. He'd seen the faces of many Calusa warriors in his sleep already. Faces, he told that were angry, faces with animal eyes, faces, burnished in war paint that would frighten a foul hearted bear!

Clugar looked aside to see Michea lost in deep thought. Something was troubling him Clugar knew. There could only be two things that troubled Michea and that was being too greatly outnumbered in battle …the loss of Lichea, or of course, both. However, Clugar not wanting to open up an old wound over the subject of Lichea's death naturally sought to avoid it.

"You are troubled over next Calusa raid?" asked Clugar cautiously.

"No," Michea replied stiffly.

Clugar seemed relieved, knowing the next battle, was of no concern to Michea. He was simply missing Lichea. Yes, Clugar knew how Michea felt for Lichea and it hurt him all the more to know this too. So they both sat with their sorrows, faces lit by the fire.

Sometime a while later, maybe thirty minutes or so, Michea seemed to snap out of his fireside trance. "Everyone must listen to what I say," Michea said with great authority. "The ember of this fire is now spewing the sparks of its wisdom. It is safe to leave it. Follow me to where we will rest for the night."

Without question, seven warriors, Clugar, and Michea, searched into the darkness for a place far away from the fire. There was something about this decision Michea made. It was not a popular decision, but no doubt, one they would find to be a prudent one. It seemed the three cowardly Calusa Warriors who took flight from battle, suddenly grew some courage. They circled back, lured by the light of the campfire, knowledgeable with the idea that they could slay their enemy while they slept. Michea never slept when he was on a mission to protect the settlement. Besides, he knew how battle tactics change. Bad tactics meant unfortunate outcome and how it could all play out if he weren't here to put a stop to it.

There were only three of them, he knew. They were cowardly warriors, certainly not worthy for his men to worry about. Let them sleep. He could handle these jackals himself. He had his silent weapon of death in each hand, an obsidian sabre and knife. Such tools should make quiet work of these warriors. He took with him one arrow and bow. With that, he nocked his arrow making it ready for the first of the three scavengers to reach his old camp site fire. Michea was close to this one, but the Calusa did not see him. As the warrior made a hand motion for his other two warriors to bring up the rear, Michea quietly lurched out of his grassy hiding spot. Holding his bow in his right hand, arrow lightly nocked in place, he came up with his left hand, obsidian knife poised. A quick sweep of his obsidian knife opened a gaping rift in the neck of the lead Calusa Warrior. The knife easily ripped the warrior's neck with deadly precision and Calusa blood ran hot down Michea's forearm. He felt the knife nick the tip of his foot as he dropped it to make quickly the transformation to a bowman. His right hand making home his pull on the nocked arrow, the feathers brushing his face as he pulled hard, drawing the arrow for a full-hilt mark on the second Warrior dashing forward to meet his comrade, meeting Michea instead. The arrow took its mark before Michea could design his best idea for a speedy kill. Unfortunately, it wouldn't be a quick kill and it would serve to cost Michea time he needed for the third and final attacker.

Michea's arrow entered the second Warriors mouth with the arrow tip smashing hard against this man's spinal disk barely splitting it enough to sever the spinal cord. Sufficient spinal activity, kept the warrior solvent and deadly … still.

There was the sword, but it lay in the darkness somewhere afoot. Michea could only in the short time, dispatch this fighting dead man as quick as he could before the third warrior, sinks a stone hammer in his skull. His heart pounded and he knew this may well be his last stand. Clearly, these men made a pact to come back and kill as many Ponctoan warriors as imaginable. They were far from being cowards. They were to be admired Michea thought. They had a war plan and although their plan did not go well, the alternate part was being implemented now. Michea knew the arrow shot on the second warrior went wrong. He was going for a chest shot and instead a lower head shot resulted. This Warrior will be dead long before the sun comes up, but Michea did not have the luxury of waiting until then. Luckily, this warrior had no strength left to make his stone hammer swing on target and Michea snapped the rest of his neck free from his head in the momentum of the pace of battle. It was a bloody and messy moment, Michea's torso smeared entirely in enemy blood.

The third and final Calusa Warrior leaped forward out of nowhere to his right side. Michea's instinct to drop suddenly may have very well saved his life. The Warrior had a long war dagger made from sharpened deer antler. Not as deadly sharp as obsidian, but it could rip open a belly or a rib cage just the same. Being disembowelled by such a tool was a Calusa pastime. They were practitioners of the art of disembowelment. And they could do it on the 'fly' and in the heat of battle! But luckily for Michea, this warrior did not have a good grip on his butchering knife. The man had a peculiar lankiness about him. Perhaps his awkwardness was a result of having grown-up too fast. His long arms made the swinging arc of his blade inaccurate and his knife penetrated Michea's right side with a curious 'pop' ..."

Michea had the back of this one's head under his chin and no weapon to sever his spinal cord. He was wiry and a good head taller than Michea. Nevertheless, the warrior had signaled his folly. The mistake that Michea needed to put him down for the kill. The large warrior reached up and forward to grapple with Michea's head. Michea's right hand dropped to retract the war knife from his side. A loathsome gurgle sound escaped with the knife as he pulled it free of his gut. Locking the big man's head tight under his left arm, Michea stabbed into the Calusa's neck vertebrate several times, until he felt the man's sides stop heaving. Effectively he had paralyzed this warrior and he let him fall into the fire. Michea tossed the butchering knife into the fire with the warrior, who was wide eyed and crackling in the embers. "Sleep warmly my brave enemy," Michea said.

Michea eased himself down to rest on his left side. He saw the wound his obsidian knife made on his right foot. In the heat of the battle, he had dropped his knife and it nicked his little toe, taking a small share of flesh the size of a baby's tooth.

The wound to his side was both painful and puzzling to Michea. The pain took a sizable stroke of his lung function, but not in the sense that the butchering knife wounded his lungs. The knife entered his right pelvic hip area, at which point deflecting over the length of his pelvic bone and slightly exiting out his right buttock. The wound burned like fire now and Michea found it difficult to breathe deeply.

Spanish moss was a favorite item used for fire kindling and medicine. The small pile of kindling near the fire was reserved for a morning wakeup fire. Michea made the exception to those plans by taking all the Spanish moss he could, so a plaster could be made and applied to his wound. Scraps of palmetto cabbage heart and a hot red oak poker made a field dressing cauterization, a painful but necessary procedure. A slight pressure applied to the wound, using a thin deer hide strap, stopped the bleeding and even some of the pain. This makeshift field dressing would serve as a bandage until he could make it back to the settlement. And that journey would have to wait until sun-up. Up to this point, breathing came easier and the pain all but became a dull ache. Michea in the warmth of the embers nearby slowly made his distance from the fire before dropping off to sleep. He was too weak right now to pull the dead warrior out of the fire. Soon, he knew this warrior would cause the fire to flare up and create a dangerous burst of intense heat.

Michea felt an enormous weight of sleep come over him. And for the first time in a long while, he dreamed. It was so vivid to him that his whole being shone through him in such a way that those in his dream new him as he appeared to them.

In his dream, Michea walked along a path of red sandy clay. At his feet, a deadly snake crossed over his wounded foot without striking him. The snake merely looked back as he departed from him. And he realized the snake was laughing. Something dreadful happened in this place; something beyond his understanding. While his attention was focused down at this strange event, a voice startled Michea. "Be not afraid," the voice said. "You have been spared the sword!" And as the Sword was fully retracted, he closed the distance between them as if to float across the pathway to Michea's side. His injury pained him with a fire that subsided as quickly as it flared up. The Sword merely brushed his side, as if this glorious warrior was testing Michea's wound for pain. Even so, Michea's attention was taken up by this one's magnificent Sword!

And what a Sword it was; Michea never seen the likes of a weapon such as this. It glistened and shone like the still waters of a crystal clear stream. In it, he could see his reflection. And the warrior who wielded it was a mighty fine figure of a man. He was dressed in exquisitely woven fleece and his whole being shown like he was in the rays of a beautifully adorned morning sun.

"Who are you," Michea asked.

"Former guard of this place," he said sadly, "now, I come here as a caretaker of this garden." Despite the inclination to look away from the face of this glowing Angel Warrior, Michea knew that would be a sign of weakness. "You handle yourself quite well," the caretaker continued. He then extended his sword out slowly with his right hand. "You will take my sword and dispatch the snake?"

Had this so-called caretaker seen him in battle, Michea began to wonder? *How well does he know how I handle myself?* Confused, he suddenly realized this beautifully adorned Warrior was awaiting his answer. He looked to see the snake still loitering in the grass as if to make fun of this decision. "I have no need of a snake," Michea said. "I am not hungry!"

The snake stopped laughing. Michea looked back to the Warrior who seemed pleased at Michea's reply. Suddenly the snake 'hissed' and slithered away!

Out of this test, the Warrior knew Michea saw the snake for as it appeared. It was just a snake. Even when it laughed it was just a snake to him. So it was, Michea's ancestors were the indigenous ones to this place! With this, the Warrior gave this warning, "Not all things are what they appear to be, as in this place you see behind me!"

"But it is a beautiful place," Michea began, "How could it be not?"

The Warrior explained that following all that is good comes evil. As it is certain, where rabbits dwell fox will surely follow. So if there is to be a balance in the end, there is no need for a protector in the interim, such as himself.

Michea was young and lo his heart pure, "Be as you are," the Angel Warrior said, "and I will be at your side."

Michea felt fortunate in that this one would offer to be his spirit guide and protector. But perhaps there was more ….

"There are many good men among my people," Michea announced, "that could make love of this land! We could stamp-out the evil bad ones in time and once again this place will be as it was!"

The Warrior transfixed his lips with the finger of his hand. "You of all people should not speak of such!"

Michea hung his head as if a scolded boy. An unspoken thought reached his mind. *My people have known of this place.* And yes it was true as the Warrior had just now related that the Ponctoan had seen this place and this place had seen them.

"What do you mean Great Warrior? Has this place eyes?"

He only nodded his head.

"Where are these eyes?"

The Warrior swept his hand over the horizon of the land behind him. Then turned to Michea and said, "In there is a tree that marks the beginning of all man. It saw the innocence of your people fall by the very fruit of this tree."

In an innocent way, it all made sense to Michea. It was his parents who taught him of all the many plants that were poison and could kill him. Learning of these things, would have meant, that for some of the first people to have tried such fruits and plants, they died as a result. He shuttered to think of how many may have died, just to learn of these things. Nevertheless, if what this Great Warrior was saying was so, then it was only the one who discovered this tree, who did the dying. So how could a single tree be as significant as to drive away his entire Indian Nation, or, as it appeared, everyone else? *Except snakes*, he mused to himself. Michea knew the names of many poisonous plants and fruits. "So tell me Great Warrior, the name of this tree."

"It was the Tree of Life, otherwise the tree of knowledge."

Didn't recall the name, so Michea knew it must be very rare, except for in this strange land of landscaped hills, trees and flowers. "So these trees are plentiful?"

The Warrior simply swayed his head with a frown, "There is only one," he said woefully. And its roots grow long."

Michea did not like this. A fearless Warrior to speak this way could only be borne of treachery! He must have seen something very wrong or very horrible. Or both! "You say this tree was the Tree of Life and you say it as if it was?"

"It transformed one day into the Tree of Death," the Warrior said as if a warning to anyone who dares passage into the forest. "It will kill man or animal that comes close to it. Seeing this tree is by the most part too late. So be warned my friend!" The Warriors face turned grim.

Michea's heart pounded "But knowing this I should not worry. There is nothing here to dare coming back for!"

The Angel Warrior's laughter thundered all around Michea. "Oh, but yes there is! When you become lost, you will be tempted to do a most dangerous thing. For the answer to your way home is in the boughs of the Tree of Death!"

Michea, puzzled by this, went on to surmise that this held a promise of some kind in the future scheme of things. This would a feat of magic, he knew. "And you will be here to help me?"

The Warrior smiled. "We only see each other for the time were right in your spirit to do so. In the alternate of time, it will not be so --"

Suddenly, the sounds of sawgrass parting with a loud 'swoosh' startled Michea. His eyes opened as the impulse to leap to his feet overtook him. The dull ache of his side, only a reminder of last night's brawl and the knife wound he sustained out of it all. A fleeting lightness in his heart left his whole being refreshed as if he had taken on a new energy. One of which he could not explain. Surrounding him now were his friends whom discovered his absence and came looking for him. How long had he slept; he wondered. And the confused looks of his warrior brothers added to the mix of unanswered questions that followed.

All around them were signs of a great battle. A battle that went forward to the campfire they had left smoldering the night before. It now was blazing and had triggered a panic among the small group of warriors who knew this should not be happening. When the fire was left, it still had many embers concealed under a thick blanket of ash. Enough ash to keep it insulated from reigniting. That is, unless someone stoked the fire with another layer of dried wood.

A great sigh of relief when Clugar found his friend Michea not missing at all. There he stood wild eyed as if he were ready to fight ... and fight he did? Clugar just noticed the two dead Calusa Warriors strewn along the path west of the campfire. Then, spotting a third one lying in the fire, all charred and blazing away, he said, "See you spent the rest of the night by the warmth of the fire!" Now it was pretty obvious why the fire reignited itself. "Ah, a Calusa warrior helps to make a fire last until morning!"

Clugar gasped at the sight of Michea. As he drew closer to the fire, Clugar saw Michea nearly covered in dried blood. In the light of the fire, Clugar also noticed the bindings that could only mean a mortal wound. Michea said nothing at first. He looked as if he were still in some kind of a trance. Clugar quickly moved to Michea's aid. "Here," Clugar begged. "Sit."

Not knowing how far along his injuries had gone over the past few hours; Michea knew he was in trouble. Odd, his pain seemed all but gone! Could it be the dream of this spiritual place and the swipe of a magic sword? He knew the Chief or the Doctor could answer his questions. There was a high order level to reach in order to become a medicine doctor. *Yes*, Michea thought. Ciabee would tell him and administer to his wounds. Hopefully, not to administer death as such a wound would decide his fate or fortune if it so be.

"No," Michea protested. "We will start for the village now."

It was a good day's journey. Clugar summoned his warriors to ready the task of carrying Michea back to the village. Again, Michea protested. He would waste no time and walk the way back on his own. It worried Clugar at first, but soon as the day drew toward noon and all became a bit hungry, it was Michea who ordered them all to stop and take a break. It was a good sign to see Michea eat. He took water and a handful of an avocado rind for a snack. And while resting, he checked his wound. It was a good sign to be seen, one that took Michea by total surprise. There was very little blood, revealing a nicely healing wound, which looked a month old, instead of hours old. This filled Michea with a renewed feeling of strength. He asked to continue and they reached the village sooner than expected.

Clugar had heard the entire account of Michea's battle with the three returning Calusa warriors. While Clugar knew the risk of sleeping around the campfire site was real, the idea of sneaking back to the site for a possible reprisal of the battle never occurred to him. He and his men felt safe in the remote darkness of the forest several hundred feet to the east. It was just the 'boy' in Michea that hungered for a good fight that lured him back to the campfire.

"You know the Calusa renegades would have broken off and went home once they found we had left," Clugar reminded Michea.

"The chance to send three more Calusa Warriors to the animal side of life," Michea began, "means three less to have to contend with in thirty or so days from now!"

No one could argue with truth and Clugar laughed. This young man was wise before his time, he thought. "You may one day become my Chief!" Clugar exclaimed. "But you must live long enough till that day comes!"

The message that Clugar was sending was well meant. He saw honor in Michea's daring decision to double back and defend his brothers. On the other hand, his decision should have contained the wisdom of all the warriors present in the war party.

This discussion grew larger in the days to follow. The Medicine Doctor also had his thoughts on this very same subject and levied his advice on the brow of Michea. And yet at the same time, he gave blessing for Michea to return to battle. Most importantly for Michea was the Doctors thoughts of what meaning his dream had and how this Angel Warrior healed him.

"You have seen the place of our beginnings," Ciabee, the Medicine Doctor told Michea. "And that place exists, but it is in strict silence of which I cannot speak. We must never share of its location."

"You know of where this Great Garden is?"

"Yes"

"Tell me Great One!"

Ciabee sighed with a painful look on his face. His head lowered as if remorse filled his heart. "It is a curse that, I like our forefathers, should know of the place we were driven from."

Something he still didn't understand made him feel bad for the Doctor. But what could it be? He could now see it in the Doctors eyes. It was a look of shame! "What have we done?" Michea whispered.

"We befriended evil," the Doctor replied. "And now we shall be forever punished." Ciabee took out the Ponctoan Spear-point and held it before Michea. "This you see has been sent us by god to redeem us!"

Michea had not told the Doctor everything he had experienced in his dream. Although it became obvious, the Doctor knew some of those things. There were many stories passed by word of mouth over the centuries to successive leaders of the Ponctoan Tribe. Ciabee had told Michea that by being visited by the spirit guide in his sleep and leaving him healed in spite of a mortal wound was proof that Michea would be his successor. His talk of this place; a garden in the jungle would now take shape in secret conversations they would share in the future. Ciabee wanted only to know that Michea understands his place in this matter and never speaks another word of it to the people of the tribe.

As Michea departed through the doorway of Ciabee's hut, Ciabee knew as long as Michea had these dreams while being injured, he would never die.

Ciabee had seen the signs years before the spear point came to them. But now he feared the inevitable. It would be the day when all return to the Garden. Although entry into the Garden was forbidden, it became clear it had to be done. A pilgrimage to this place would soon be necessary to preserve humanity from imploding on itself.

Yes, it was obvious to Ciabee now that if the Ponctoan Spear fell into the hands of those evil doers like Max Tanker, the secrets of the past may become accessible. Innocent lives would be destroyed. Ciabee was still strong, but coming up in the years made him notice he wasn't getting stronger. The day soon will follow when he must conceal this deadly icon. Ciabee shuttered at the thought of this place.

In the days to follow, Ciabee came to believe it important for Michea to see the battle he was about to engage in. It would mean Michea would have to go into the future together with Ciabee. How this could be done would come down to experiment. As before, in the case of taking the former maiden Chickpea into the *Spirit-side*, he carried her in his arms. This would not be possible for one as big as Michea.

First, Ciabee would rip a hole and enter grabbing Michea's hand so that he could follow in close behind. This attempt failed, as Michea only felt Ciabee's hand melt away, before he could follow in. It was like everything went flat just as Ciabee slid through the hole and disappeared.

Michea tried to describe what he saw the moment Ciabee disappeared. "You become flat as sawgrass," he said to Ciabee. "Then you become motionless, yet you rapidly moved from sight into nothingness." Michea was scratching his head, for it seemed, nothing he could say would make sense of this transformation into the time-continuum for which Ciabee embarked.

Time was slipping away. Only a few weeks remained before the Calusa War Party would come. And Ciabee was out of ideas. It became obvious that Michea would have to go in to the time-continuum alone.

His finest Warrior, the Prince Warrior of the Ponctoan, otherwise known as Clugar was one of the bravest warriors Ciabee had ever seen. Clugar was fierce in battle and had a record of tracking down his enemy. Michea on the other hand, was shaping up to being one of the best strategists Ciabee had ever seen. This was why he needed Michea to see this battle. There were things of great concern to Ciabee that maybe Michea would see and solve long beforehand. One thing was for certain if anything, evacuation may be the best strategy in either case. Ciabee had already prepared for this battle. Arrow makers were busy making arrows in abundant supply. All war arrows and some game arrows were converted at the notch. The Chief had agreed that if the battle came from two sides of the village, seven of the village's best archers would greet them, high up in the trees!

Ciabee explained to Michea all about time travel. He went over all the fine details of what Ciabee himself knew. "Above all," said Ciabee, "you must watch your step. You will be following a stone path that crests atop an infinitely high wall. If you fall or step away from the path, you will fall into the time you see before you. Getting back to this time is easy, but you must remember the exact place you came."

Michea knew of the oddity of the stone path and how the past crumbles away. Just how fast that happens, had to be Ciabee's guess. In that he surmised that for every moon phase, a day would lapse and crumble into infinity. "I will not be gone, but for whatever it takes to see what I must," he told Ciabee.

"Look at, but do not engage this battle," warned Ciabee. "To enter the near future would not be wise!"

Michea looked confused. "What would happen Great One?"

Ciabee shook his head and pursed his lips. "Something dreadful will happen that cannot be undone. You will have entered into a fold of time that overlaps your own existence. You can only observe. To re-enter you must re-trace your steps back to the path that ends and allow yourself to slip back as the path crumbles beneath your feet!"

Maybe it was just a guess, but Ciabee seemed to know plenty on this subject. To dare to ask how it was known seemed a mockery that Michea had no words for. He kept himself silent on this, feeling safe and secure in the thought that what the Doctor says is what is.

It was late that evening as Ciabee prepared Michea for his first jaunt into the time-continuum. Several times Ciabee went over what Michea needed to know before making his entry point into the unknown fabric of time. Guard at all times the spear point Ciabee warned. "Never let it get away from you … ever!" In addition, Ciabee said sternly, "You have your people in the palm of your hand," he added as he passed the spear point to Michea's hand. "The concern over your own life pales to that of all your people."

Ciabee knew this was already ingrained in Michea's soul. But when the spear point was placed in his hand, Michea knew of nothing but the value of his people over his own life. The power of this spear point flowed throughout his body, with an energy unsurpassed by that of life itself. As if all his people and their very being was throbbing within his blood.

Michea knelt before Ciabee. He looked down at the spear point in his hand and swore to his Medicine Doctor that he would not fail. This, he knew was the very first time the spear point had been transferred to anyone but the Chief or the hand of the Ponctoan Medicine Warrior, Ciabee the Doctor of the tribe of the Ponctoan Society. It was a great honor indeed! He would guard this powerful icon with the last breath of his life!

Just like Ciabee described, Michea chose for himself a small black hole and with the spear point, he gouged the hole and ripped out for himself an entry into the time-continuum.

Ciabee stood looking on. He saw faintly the rip Michea made with the crude stone tip. And now he realized what Michea meant when he observed himself become a still and flattened image. And then by an immeasurable stroke of speed that image seemed to be sucked through the rip in the fabric of time.

Now Ciabee stood by the doorway to the Ceremonial House waiting for Michea's return. A few more minutes passed and yet, no sight of Michea. Ciabee drew a worried breath and slowly sank to the steps of the Ceremonial House. As the sun began to set, Ciabee stood worried that trouble may have come to Michea. There would be no sleep tonight, knowing of what Michea may see on the other side. Ciabee witnessed scenes of total chaos and finally, annihilation of his people. However, at least now having knowledge of this, Ciabee had a partial grip on making a good stand against the attackers. Still, Ciabee worried, even though he knew Michea was trustworthy and strong. But anything could happen.

Chapter 11:

Michea's arrival at Davistown Pennsylvania.

Ciabee gave Michea a pretty good description of what to expect when entering the time-continuum. Immediately, he felt the stones at his feet begin to separate and crumble out. He took a few steps forward and watched as the pathway slowly crumbled away. And as expected, the crumbling stones progressed along the path toward him. If he stood long enough he would slip back home to his village. It was strange to see, but it was a good thought Michea believed.

As it came time to move ahead a few weeks, Michea walked a few steps and then stopped to observe the images of his village. Hard to tell of what day it was, but it appeared to be early in the day. He saw his mother dump some wash water from a pot outside their hut. A chore she did for many years. Michea smiled, it meant the morning regiment to help begin the day by washing and breakfast preparations.

Michea moved on. He heard the crackle of thunder far off down the path and an occasional snap of lightning. Voices could also be heard but too low to make out what was being said. There was a cacophony of sounds, both familiar and unfamiliar. There was a voice seemingly well known, but far along the path it was. It was a familiar voice yelling out a name and Michea, for the moment, knew this voice.

Before long, Michea stood over many images that looked to be far into the future. Here was a place unfamiliar to him, with people in soft green clothing. Some wore other colorful clothes made of cloth that bore patterns of beautifully arranged colors. More and more people rushed into this already crowded room. Someone peered through a doorway, somebody familiar to Michea. *Was it the Doctor of his tribe?*

Michea could not be certain. The figure was only there briefly before being pulled away by someone of *this* tribe of colorful people. Michea jogged back to replay this incident, over and over. It was Ciabee for sure. *He was here, but why?*

Michea broke into a run. Forward down the path, stopping briefly a few times to view what all he could see and then running some more. Finally, there was a view of a small figure in a bed. It was a young girl wearing a gown and a baggy hat covering long black hair. Michea gasped. On the bedside table, lay a small straw doll. The child slept while little beeping sounds could be heard faintly around the room. There were many vine-like appendages attached to this child. A man appeared at the doorway to this child's room. He appeared eager to enter, but there were womenfolk in colored dress pushing him pack. A voice sternly said, "No flash pictures!" Michea drew as close as he could. Trying to make out what these words meant and what this peculiar man in a dark cloak with the strange bonnet on his head. He squeezed around the woman and held up a black short barrelled weapon? Attached to it was a silver bowl.

A sharp pain pierced his heart. *Was this young girl, his wife Lichea? Is this where the Doctor took her? Where was he now?* All Michea knew was this child was in mortal danger from this man who came to finish her off. Michea was ready to lurch in and save the little one, whom he was nearly certain had to be Lichea.

But it would never happen.

The light of a flashbulb nearly burned his face, or it seemed. He was so close the image of the silver bowl was all he could see. The muscles of his back nearly snapped his own spine from the involuntary action of this surprise lightning bolt that seemed to have torn into his eyes. His training in battle taught him to gain his footing quickly when knocked down. He remained fluid and established a quick defense. By his left hand, he felt the edge of the walkway and in his right hand, the spear point. He was up on his feet, but feeling very disoriented. His eyes wide open and all he could see was a bright spot that blinded him from all sight of where exactly he stood. The breath he pulled was of a sweet pungent odor that burned the passages of his nose. Struggling to recover his senses, Michea fought to gain a sense of the general direction of the assault. However, he was already dangerously close to the edge opposite of the images he had been viewing.

Michea lost his balance when he tried to correct his stance, but discovering he could not recover solid footing; he slid off the opposite side from where he saw Lichea. He landed in Davistown Pennsylvania. That January night, 1966 was cold and Michea, being a native Floridian, awoke from having struck his head, in a fit of convulsions. The back of his head felt like it was on fire and someone stomped out the flames with a war club. A dark figure loomed over him. The figure silently prodded him with their hand or maybe a stick, Michea wasn't sure. One thing became quickly apparent to him. This stranger had slapped his face.

"Guys," the voice said loudly to others in the darkness behind him. "It's just a drunk beatnik or something … let's roll'em and see what else he's got!"

Michea was up on his feet all before the stranger landed on his ass. Michea exchanged footing with the stranger, tripping him and bringing him down. Now for the first time he could see well enough in the night to see the faces of others closing in on the two. Obviously, the brotherhood of the warrior he just tagged. He was in hostile territory and he would have run to escape and to keep warm. He took in the surroundings of this area as he knew he would have to return to this place to escape back to his village. There were a big orange and blue sign with the symbols RX on it. It was on a corner of travel ways that machines made way. He ducked into a doorway. It seemed to be the perfect cove to take shelter from the wind. He watched the machines and their bright eyes travel in disciplined lines along the hard beaten path.

His mind wandered back to the escape. He strained to see if any of them were giving chase. All was quiet. Perhaps it was a good time to go back and re-enter the time-continuum. All he needed was a short bit of time and he'd be off on his own once again. Slowly he began his walk back. But both hands were empty. What did he do with the spear point? His heart pounded as he checked his pouch. There was only the assortment of arrowheads, plus the beautifully crafted obsidian arrowhead his Mother made him for good luck, but not the Ponctoan Spear. It had to be lying on the sidewalk. Michea broke into a full run and he did not stop until he reached the corner with the sign RX. He could see the tracks in the snow. He also saw his own imprint in the scatterings of the snow and ice on the sidewalk.

Michea fell to his knees. He searched with his hands to areas where the spear point may have slid. *It had to be here!* There was no way it could have been left on the time-continuum pathway. Nevertheless, it was a point to terrorize Michea's mind leaving him in a state of conceptual disarray. Despite this state of mental shock, Michea knew he clutched the spear point as he fell. He was certain. It had to be here. Again, he searched. Over and over until his hands could stand no more of the frozen punishment, he had put them through. Minutes later he found himself shivering and looking about frantically for a place to go, a place where at least he could warm up. He saw a door swing open to a stone lodge across the street. A man had just left while another was entering.

Michea was there in a flash and followed in behind the stranger. Keeping close to the door, Michea took immediate comfort in the warmth within this place. The smell of food crossed over his nostrils and the clatter of dishes and smoke filled the air. In a lot of ways, this was like being home. There were two men standing near a beautifully adorned table with slender spears in their hands. One slid past the other, bent over the table and with a graceful stroke of his slender spear, struck a stone ball that clattered into other stone balls. All at once, there was laughter from a dark corner of the room. Tucked away around a table set within a nook, was a woman maybe two and two men. Something was funny and they all laughed. Another woman swept past Michea with a table board in her hand. Shimmering columns of amber drink sparkled from within. She balanced those columns as gracefully as she walked, that Michea could barely take his eyes away from her. Along came a woman dressed just like the other and approached Michea with a smile.

"Would you like to take a seat," she said. Michea understood Spanish and could speak it reasonably well. However he understood English far better than he could speak it. Which was okay for it would serve him well enough, so long as he didn't get into a debate with anyone here. He smiled and shook his head. "Maybe you would like to sit at the bar?"

"Si," Michea nodded.

Once at the bar, Michea became mystified by the pretty beer decorations and neon signs. This was a very nice and friendly place he thought. The man next to him was brought a tall mug of draft beer. Michea watched to see how the bartering would go. The man slapped a few coins on the bar. They were bright shiny silver pieces. Michea silently marveled. There was no arguing, the bartender picked up the silver pieces and deposited them in a music box that made a 'cha-ching' when it opened. Then the tall lean man with the white apron walked over to Michea. He gave Michea a good 'eye washing' and with a glint in his eye, he smiled and asked, "What can I get fer ya Geronimo?"

"Name not Jerome. Name is Michea, I am here for warmth. Be on my way now!"

"Wait a minute Mickey," said the bartender. "Coffee is free here. I will give you cup if you like!"

"Let me buy him a draft," said the stranger seated next to him.

The bartender turned and on his way to the coffee urn, whispered into the stranger's ear, "We don't serve liquor to kids." The stranger looked hard at Michea, then reached into the vest of his jacket and pulled out a pair of glasses. Michea turned and glanced his way as he placed the glasses on his face. The stranger's eyes bloomed under the magnification. And with that, so did Michea's eyes opening large. "IEEE," Michea shrieked.

Silence fell over the tavern. In that brief moment a dish was heard breaking way off in the kitchen.

The stool Michea sat on tipped over as he leaped from his seat. So startled was he, that the stranger too, became startled. "Sheesh kid!" the stranger exclaimed. "You scared the shit outta me!" He then took his glasses off seeing Michea looking shaken and confused. "Are you going to be alright?"

Michea felt the paint of embarrassment covering his face. He bowed his face in humility and held up his hand. "I beg ... forgiveness." He knelt down and resurrected his stool to the bar where it belonged. A mug of coffee made a 'clunk' on the bar in front of him. Slowly he lifted his face to see the smiling face of the bartender.

"You do that again," said the bartender with that smirk on his face and pointing to a shallow basket of nuts. "I'll have you grinding coffee beans with that nut-cracker over there!"

Had the house not been so silent, those words would have never been heard but for Michea's ears. With that the place roared with laughter. Michea didn't quite understand it all, but the tone was jovial and he joined-in on the laughter. After all, he thought, it was a funny moment, with no harm done.

The coffee smelled heavenly, though the taste was not as good. It was hot and oddly, tasted better with each sip. One of the womenfolk which lived here passed between him and the stranger. Michea loved the way she smelled. She moved close to him and looked at him directly in the face! Michea was not certain if he should allow a strange female to stare him down! However, her eyes were soft and sympathetic. "Listen," she said. "I saw that wrinkled nose when you took a sip of the coffee." Michea wasn't sure what she had said. *She talked very fast.* "Here," she continued after waiting for him to reply. "Try some of this." She reached for the little white pitcher and poured a good draw of creamer into his coffee. She took his spoon from his saucer and stirred it for him. "There. Try it now!" She stood waiting for him to try it and Michea was pretty certain she would not be pleased until he showed his appreciation. So Michea took a sip. *Much better,* he thought. He looked at her and gave her a nod and a smile. When she turned and was a good distance away from Michea, the stranger leaned toward Michea and slyly told him that the woman could '*mother him*' any day! Michea could only smile back at the stranger. *She was much too young to possibly be his Mother, unless something went horribly wrong!* Michea pondered this. His coffee tasted good now. *Perhaps this man made a bad turn in this time travels, like himself. And now he is older than his Mother.* The thought made Michea frown.

"I hurt for you," Michea said earnestly, "and your Mother is very nice."

The stranger chuckled and shook his head. "You have got to be older than you look. Lemme guess," The stranger rubbed his chin. "Twenty five, right?"

"Fifteen maybe sixteen seasons."

The stranger laughed it off saying, "Yeah and I'm twenty five," when obvious to Michea this stranger was much older.

"You hold together well," Michea said politely. "But you say you should be twenty five I see," while he pointed to the woman over his shoulder as the stranger's Mother. Or at least Michea believed her to be thirty five years old ... maybe.

She ducked her head between the two. "I see you boys are enjoying yourselves!"

"Yes," Michea said politely while flipping his thumb over his shoulder to the stranger. "Your son is a good one."

"Yeah right!" she said sarcastically. Then she gave him a razz-berry! Michea got a face full of saliva. He gave her a broad smile

He felt honored.

Chapter12:

Around the corner trouble lies

Michea found it was time to go home. The tavern was closing and time was slipping off. Sunrise was still hours away and Michea would not leave this area until he found the spear point. He remained vigilant huddling in the cold inside the alcove doorway of the drug store across from the tavern. He managed an hour of sleep. Still, it was night now, with only a car passing every now and then. Just as soon as the light of the sun arrived, Michea would find the spear point. It had to be lying in the snow that clustered in some spots along the sidewalk. He began to think that after he had hit the back of his head on the sidewalk that he may have blacked out briefly. ***There were words spoken.***

What were those words and who were all those people?

Michea struggled hard to remember and although some of those words he recalled, he did not know what *'hippy'* was. He mouthed the words several times. He needed help to know what these words were. In the tribal order of things, there were no such teachings of these words. In this world, words were everywhere, printed on signs and buildings. Even so, as the sun began to rise, Michea was awestruck at the size of this enormous settlement. Colossal fortresses made of stone lined the thoroughfare. Even windows were sealed shut with a clear stone so that one could still see outside. And for the first time he realized automobiles carried people inside!

And now the time came. He would reach back into the training he received in his world. A skill that instinctively marked Ponctoan Indians, passed down from generations long before Michea was born. It was the talent of reading tracks of both man and animal. Michea dashed around the corner and in the first light of the morning sun, he studied the tracks left behind from last night. He could plainly see where he had been and where he had come back to search for the spear point. Everything was there, clearly visible. However, a place in the snow close to where the sidewalk met the building laid a print of the back of his right hand. This, Michea knew was the original landing point of his hand when he fell to the sidewalk. From his hand print, he could see an additional line leading off from where the crest of his index finger knuckle met his palm. It was all Michea needed to know at this point.

He blew a sigh of relief. The spear point was here. It was in this world with him and not in the *spirit-side* of the world. Michea worked as fast as he could to clear the sidewalk of all loose snow, using his hands and feet. And all the while he did this, his mind was racing. He wasn't fooling himself now. He knew he had been met by bandits! It was no doubt a roaming bunch of bandits who may or may not frequent this area of the settlement. Nevertheless, if luck would have it, they would be here again. Bandits were a different kind of hunter, but their tactics were very much the same. Their game trails, instead of animal consisted of man!

Suddenly, with the sound of scraping behind him, Michea spun around to see an old man working to clear the sidewalk with a tool. He had a thick jacket and furry hat with ear flaps. "Hey," he yelled to Michea, "that isn't any way to clear a sidewalk." The old gentlemen handed Michea the ice scraper and then disappeared around the corner. Soon he returned with another ice scraper and two small bags of salt. The old gentleman pointed to a trail of icy footprints on the sidewalk.

"You see those," he asked Michea. "That's Mrs Smith's footprints."

Michea ducked down and carefully felt the profile of one of the footprints. He then gently moved his hand over the top of the footprint taking in the ridges left behind by the sole of her boot. The old gentleman only frowned and shook his head. "It just doesn't matter how early I get out here to clear the sidewalk, she'll tromp across the snow causing ice to form." He stopped with a sigh. He just needed to get this complaint off his chest. "But we don't need to try to scrape them all; we can use this salt to do the job!"

Michea dropped his scraper and traced this little boot prints all the way by where he had been last night. The prints looked as if they had definite continuity; no sign of pausing or stopping. Obviously, Mrs Smith did not find the spear point. It was gone before she got there. Michea did not have the key knowledge of snow tracks to tell him how old the tracks were, but he knew that they had to have been laid in the snow only a few hours ago. Perhaps during the time he crouched in the alcove of the doorway around the corner. She walked right past Michea without his seeing her!

The old gentleman only stood staring at Michea's every move. Puzzled, he finally said, "Look Tonto, let's just toss some salt around and then we'll go inside and warm up over a cup of coffee!" He tossed one of the bags of salt for Michea to catch. Michea picked up the bag of salt, ripped it open and began copying the actions of the old gentleman who was starting at the other end of the sidewalk closest to the corner of the drug store where they stood. "Make sure you get plenty of salt on those footprints! Don't wanna lawsuit against my store ..."

It was amazing to Michea to see the footprints beginning to melt away under the sprinkling of salt. He knelt down and swirled his finger around the melting mash of slushy ice that was once Mrs Smith's boot print.

"You ready?" asked the old gentleman.

"Coffee," Michea said with great reverence and admiration for his new found hot drink.

"You bet!" smiled the old gentleman.

"You bet!" Michea mimicked.

"Sam Charles," is the name, the old gentleman said pulling off his glove to shake hands with Michea. "Appreciate all the help I can get these days!" Michea took Sam's handshake without reply. Before he could ask Michea his name, Michea just smiled and said coffee.

Once inside, Michea would not admit he was near frostbite condition. His hands and feet felt numb and when they began to warm up, Michea experienced something he had never experienced in his entire life. He thought his feet were on fire! He concealed his pain from Sam, who was busy getting coffee setup and served. And if that was not enough, Michea experienced the feeling of 'pins and needles' throughout his feet and hands. He just sat at the soda bar. Arms crossed the black marble bar top, he buried his face in his arms, trying to muffle his cry. "Waukee, "he cried. "Waukee." Michea was crying out to his Mother in pain.

Sam heard but never turned his head, "Cream and sugar, right?"

Michea knew that his condition was a result of extreme cold for which he had never experienced before. But knew the pain was now beginning to subside. He tried to smile as he lifted his head from the marble bar. "You bet," he said.

The clink of the spoon in a thick ceramic mug was music to Michea. He loved the way that sounded. And when he saw the way Sam made love to his mug of coffee, Michea quickly picked up his and took it in. It was better than last night's coffee! It was sweet and nutty. And though Sam wanted to make small talk, talking was out of the question for Michea. There was coffee to drink!

"You bet!" Michea said with a robust smile, as he pushed his empty mug across the bar to Sam.

"Ah," laughed Sam, "another cup of You Bet!" It was becoming clear to Sam that Michea was a man of few words. While Sam was making another cup, Michea disappeared out the door of the store. Michea had not relieved himself since after leaving the tavern last night. It was time to go. The biggest problem now for Michea was finding shelter for his business. It was getting brightly lit outside and people were beginning to come out from their homes. Sam's instincts were on track. You didn't live to be seventy-two years old and not gain any human instinct. He was out the door and onto Michea's trail as he rounded the corner to the back of the store. Moments later Michea was facing the porcelain fixture of relief.

"There you go," said Sam. "My store is pretty old, but not so old we don't have a pisser!" Michea looked over his shoulder sheepishly as if to say, 'is it okay to pee on something this pretty?'

Sam rolled his eyes back and shook his head as he muttered, "Never thought I'd have to potty train an Indian Chief," Sam turned muttering and heading out the door. It was the tinkling sound of the doorbell. "Customer out front," Sam said excusing himself. "Don't forget to wash your hands." Sam was strict about that since he was also the drug store's Pharmacist.

By the time Michea came out of the backroom bathroom, Sam was on the telephone talking to his wife. "Yes. Yes. Bring down some extra pants. The pea green ones I hate and those plaid shirts with the button down pockets. Hate them things and yes, those too. I'll see if I can find those boots," he said standing on his tip toes getting a look at Michea's feet. "Yea, I think they'd fit 'em. Thanks dear. See ya in a bit!"

Michea waited patiently by the bar for the invitation to sit and have coffee. And of course, Sam told him to take a seat at the far end of the bar. He brought him another cup of coffee. This time Michea was going to pay for it because he had gotten a sense of bartering from last night. *It was time to begin paying his way,* he thought. Perhaps one bird tip arrowhead and one broad tip arrowhead would be enough. Michea slid the two relics across the black marble soda bar toward Sam.

Sam looked down at the arrowheads and then over the top rim of his reading glasses at Michea. "Where on earth did ya get these?"

"Waukee," said Michea.

"Oh, hmm," Sam said, lost in thoughtful inspection of the two arrowheads. "Milwaukee you say?"

"Mum," Michea corrected as being his Mother. "Mum Waukee."

"Yeah yeah, that's what I said." Sam tried to figure out this young man's reason for being here. That is, being a stranger and all. "What brings you here to Davistown?"

Michea knew what Sam said, but didn't quite know how to respond to that question, so he said nothing. Sam on the other hand slid the arrowheads back for Michea to take back. "For you!" protested Michea.

"Ah no," Sam rebuked. "I ain't no expert on arrowheads. You need to see a professor if you need answers!"

Michea's eyebrows lifted with surprise. Maybe this was his ticket back home? Maybe in this time period there are Doctors who know much about this spear point magic. And know how to open up the time-continuum so as to enter the *spirit-side*. However, before this can be sought out, Michea was certain to have made payment for Sam's coffee. It was a simple matter that Michea was used to doing most of his life. Michea pointed to the coffee and next pointed to himself. Then Michea pointed to Sam and then the arrowheads.

Sam caught on right away. "You don't get to be seventy-two years old and not barter for goods at one time or another ...wait a minute buster!" Sam smiled. "This ain't no game of marbles here … you're gonna work this off!"

This wasn't a problem. Michea began nodding his head knowingly following along with Sam's instructions. "Now listen," he began, "I got these customers coming in here all day long and tracking snow in the store. Snow melts and makes water. And someone's gotta mop this water up. Ya know I don't want no lawsuits!"

If there was one thing Michea loved, well, two things actually. He enjoyed hunting and working at things. Not in that particular order, but butchering an animal came as work too! And so, Michea was unprepared for the apron! "Yes, the apron. You gotta wear this," Sam began, helping Michea with the drawstring. "You'll never get the customers' respect if you ain't wearing an apron!"

Lessons, Michea thought. Yes, Michea loved lessons! Michea beamed with pride and Sam chuckled to look at him. "Ya, look like a Cigar Store Indian," Sam mused. "Don't know whether to hand ya a mop or stand ya outside the door with a decanter of cigars in your hands!"

By the end of the week, Michea learned about money. Sam paid him well each day. And each day they would go over the day's receipts counting out loud. Soon Michea seen all there was to know about money. It wasn't long before he could read the price labels on all the products in the store. The many clothes given him by Sam and his wife Sara were well received and Michea spent his overnight hours in the backroom of the drug store on a cot Sam provided him.

Soon it was learned Michea had discovered he had a talent for pencil drawings. He was fascinated by paper and pencil. Evenings were spent in the backroom under the light of a table lamp, just drawing scenes of everyday Ponctoan life as he knew it; hunting scenes, battle scenes and scenes of celebration. The one favorite and most detailed scene was that of his dream of the beautiful forest and the

Caretaker with his mighty Sword! It was Sam's favorite too and he proudly hung the pencil drawing over the soda fountain bar. Sam chose not to display a vividly graphic illustration that showed what appeared to be a man burning to death in a campfire. Sam also chose not to display a pencil sketch of an Indian transfixing the throat of another Indian, with a long spear. Michea explained that the Indian with the spear was his best friend, Clugar; saving his life while he lay at the feet of his attacker!

"Cougar?" Sam asked. "Are you sure you're from Milwaukee?"

"Clu-gar," Michea said carefully. "Mum is Waukee!"

Sam sighed and smiled. "Well, now that we got that figured out," he said sardonically, "Are you ready for soda school?"

Michea dashed around to the operator side of the bar. "Ah, lessons." He said. "Learn much today!"

But lessons were interrupted when trouble rounded the corner and entered the store. Four large youths and one small but very vocal youth marched right up to the soda bar. "Oh, yeah," Sam muttered. "It's Saturday."

Michea locked his eyes on them as they walked up. He could feel through Sam's reactions that something was wrong. The larger four had smug looks on their faces. The smaller agile youth launched himself gracefully, on to a bar stool. This brought the other four to turn a look to one another and give a giggle and a head nod. They all dressed basically the same and it was in strange black leather clothing;

Michea noticed. He could smell the animal hide and knew it to be a mammal herbivore, not like bear, but closer to deer. They were hunters Michea decided, possibly even warriors! Michea slipped a hand behind his back and pulled loose his apron strings.

The one at the bar slid his arm across the bar, all the while clicking open a switchblade knife. "Hey Pop," the young man snickered. Pausing to point his switchblade toward Michea, he asked, "Who's the hippy stooge?" The look on the young man's face changed to a quizzical expression when he asked Sam, "Look at the moron washing his hand in your cherry sauce!"

And that Michea did. In the row of stainless containers at the edge of the bar were many flavoured sauces. Michea had a 'dead-eye' stare into the face of the young man while his hand lay idly immersed in maraschino cherry sauce. When the two locked stares, it was as if Michea hypnotized the youth with his hauntingly black eyes. Then as swift as a spitting viper, Michea lifted his hand from the cherry sauce, flinging a glob of it in the open eyes of the young man holding a knife on him. Michea's other hand clubbed the top of his opponents head while the other closely followed slapping the back of the youth's head like the paw of a bear. The young man's head hit the marble slab of the bar with a sickening thud.

The young man barely had the time to lift his head fully when Michea had whipped off his apron and slung it around the neck of his opponent. Pulling it tight he dragged the man across the bar with one hand. The man's switchblade was still spinning on the bar when Michea snatched it up with his other hand and swiftly tucked the tip of the knife in the young attacker's ear, "Michea!" Sam yelled.

Michea froze on a command.

Sam never noticed the muscles on Michea. When he was relaxed, he looked fit but nothing like this! His muscles bristled large and inflated now. "Easy Michea," Sam said slowly and clearly. "Ease him down so his feet touch the floor ... your hanging the bastard for god sakes!"

Sam put his hand to his forehead and began rubbing his brow. He was confused a bit and didn't know which way to turn. "Talk about wearing an apron to show customers who's in charge ... Sheesh!"

"See how apron works now," Michea said proudly. "Good lesson. Sheesh!"

Concentration broke when the doorbell rattled. The other four took flight at the sight of Michea and their supposed leader being hung and nearly carved up like a jack-o-lantern.

About the same time as the doorbell cresting the top of the door, Michea's attention shifted to the escaping warriors. They would go back to camp and bring many more to raid the store! Michea loosened his grip on his makeshift apron noose. The young thug folded to the floor like an old rag doll. He instructed Sam to put his foot on the thug's throat until he got back and with that, he leapt the width of the bar with the grace of an antelope and headed for the door. "Michea NO," Sam yelled!

Michea froze as he had seen the old one was wise and knew the situation here. Sam was about to yell to Michea that he was calling the police, when out in the street, he heard sirens. A street side fire alarm was pulled. The rattling bell of the alarm was barely audible, but it was the one on the pole opposite of the grocery store, on the same side of the street as the drug store. "Punks pulled the fire alarm!" Sam laughed. He had a bit of relief in his voice. "Michea," Sam ordered. "Go to your room!"

Michea looked confused, even a bit stunned.

"I said," Sam sternly, "to your room." Sam then held his hands up in resignation. "Okay, someone's going to jail and it ain't gonna be *you!*" Sam literally began pushing Michea to the backroom.

"Learn to be Soda Jerk," Michea protested.

"Later Michea … later," Sam needed time to figure out who this Indian man was, it seemed. Apparently Sam had never seen a man handle himself so effectively and precisely against an attacker. He was dangerous and extremely deadly, but Sam had seen Michea was a decent man with a good heart. Sam quickly ran around to the operator side of the soda bar. He removed the twisted apron from around the punk's neck. He was breathing and his ear was bleeding from the small incision Michea administered with the punk's own switchblade. "Want to scratch me off your list," Sam sneered. "Won't be paying you punks' a penny from this day forward," Sam could hardly hold back. He wanted to spit in this punks face, "protection money your ass! I'm gonna start charging you assholes, protection money … how's that sound you loser?"

The young man was moaning. He was in pain. Sam relished in the punk's pain and hoped he heard every word he told him. Sam sat the man up to lean against the wall behind the bar. "The cops are here," he told him. Sam, as quickly as he could move, went around to the backroom door, ducked his head in and told Michea not to come out until he said it was okay. Michea knew the danger was over, but stayed near the door just the same. He listened carefully as the fire fighters and police entered the drug store. One police officer stood over the felon cuffing him, while the other kept a watchful guard over him. Two fire fighters looked the young man over briefly. "Looks like he split his ear?" One of the two fire fighters went to work cleaning him up. The police officer in command said to go ahead. He needed to get a statement from Sam, so he could file a report.

"One last thing Mr Charles," the officer asked. "You pressing charges?"

Sam nodded his head. And with that the excitement was over. The police had their suspect ready to book into jail.

Chapter 13:

The Cage

By four 'o clock that afternoon, Michea learned how to fix malts, milk shakes, and various ice cream dishes. His favorite was a banana split, followed closely by a double fudge nut supreme. And as luck would have it, a delivery to the store came and Michea went to the backroom door to help unload, store merchandise. Luck in the sense that a visitor came through the front door, who may have taken Michea away.

He was a police detective named Henry Watson. He dressed in a manner that one would expect a detective to dress; a fedora hat, an unbuttoned overcoat that revealed his three piece suit and tightly buttoned vest over a potbellied midsection. The deep care lines etched in his forehead came with the territory and he chomped an unlit cigar with affection equal to that of a bull dog with a pork chop bone.

Sam motioned Henry to sit at the soda bar as he wiped his hands and excused himself for the backroom. Henry occupied himself by picking up a stack of Michea's pencil drawings.

Sam briefly explained to Michea that a man had come to the store to talk business with him. It was on Sam's strict order that he did not want him to leave the backroom until the 'coast was clear'.

"You bet," Michea confirmed with a smile.

Sam filled in the fact that the gentleman visitor was a police officer enforcing the law and that he was following-up on the details which happened this morning.

When he returned to the soda bar, Henry was pushing the stack of pencil drawings aside. He opened up a notepad and took out his pen. "Say uh, Mr Charles," Henry spoke. "You wouldn't happen to have pen refills would you?" Without a word, Sam hurriedly went for a pack of refills on a shelf near the comic book rack. In the meanwhile, Henry lifted himself from his stool and took a glance at the open sauce containers that lined the edge of the bar. He shook his pen and then wrote something in his notepad. Looking at his pen he muttered, "When these things quit, they quit!"

"Yeah," Sam replied sympathetic. "Nothing like the quill-tipped pens. All ya needed was to carry around a jar of indie ink and ya were always ready!"

Obviously, Henry could see Sam was being sarcastic and he returned him a slack faced stare. "Yeah, Mr Charles. Sure."

"So what's on your mind Mr Watson?"

Henry Watson was still fiddling around with his pen. "Ya gotta napkin?" Henry asked, irritated with the ink cartridge dripping ink down his hand and wrist. "The matter over the incident early this morning," he began. "I gotta follow up on these things ya know!"

"Oh by all means!"

"Especially when the kid your employee beat up is the son of a local circuit court judge!"

Sam, a bit shocked by this, rubbed his chin in thought. "Hmm," he said. "Now that would explain why I've been having to pay protection money for the past six months!"

"Whaa?" Henry responded.

"Well, ya know how these legal types are," Sam said swaggering about behind the soda bar. "The kid's gotta be a chip off the old block, wouldn't ya think?"

Henry choked, coughed and then cleared his voice, all the while looking around, as if maybe someone was listening in on the conversation. "I talked to this kid," he said in a serious tone. "He feared for his life while your clerk terrorized him!"

Sam couldn't stand for talk like this. "Listen," Sam returned, "my clerk was merely protecting this store and its patrons from this rat punk!" There were some more words, but before it was over, Sam had his face and finger drawn across the face of Henry Watson. "And another thing, if that punk ever steps foot in here again, I'll have him hung on that coat rack over there!"

"Easy. Easy Mr Charles," Henry cried. "I'm not defending this young man whatsoever. I am here only to check into this incident and write a following up investigative report. That's all. Just let me do my job." Henry Watson could see Sam was getting worked up to the point he was shaking. And Henry was not having any luck with his pen. Once he pushed the cartridge over the bayonet tube and gave it a slight squeeze, the ink still didn't flow to the tip. "Have ya got an ink blotter handy?" he asked Sam. Still irritated over Watson's bias for the Judge's son, Sam slapped down a whole stack of napkins and plucked the pen from the detective's hands. Carefully with the edge of a knife, Sam scraped dried ink out of the pen tip channel. He then gave it a slight squeeze into the napkins and ink began to flow.

"Don't look like you use this pen a whole lot," remarked Sam. "You write much do you?"

Henry's face reddened at the remark. "I use a pencil most of the time. So where is this clerk of yours?"

"He's not here right now."

The answer was too blunt and too quick to be the truth, for the likes of this detective. "Look," he said. "Let's be honest with one another here. I came to get answers. I also came to take your employee to the station for questioning. We can choose to make that happen the easy way, or we can choose to take a different path," he stopped to let it sink in. "And that path can be very ugly, I assure you Mr Charles." There was at least, compassion in Mr Henry's voice. "So Mr Charles, what ya say?"

Sam turned without saying a word. The detective was right Sam knew and his eyes grew sad as he left for the backroom. Moments later Sam was followed by a tanned skin man with jet-black hair, gathered in a ponytail.

The detective sized up Michea. He was approximately a half a head taller than the detective, about six feet tall and around 180 pounds, with dark-brown eyes. The detective, in his report, guessed Michea to be late teens, maybe early twenties. This was all a guess and was the detective's impression which matched the description injured gang member; John Gravy gave, when they questioned him. The exact description information could wait when they interrogated Michea at the station.

Michea was in for a rough night at the police station. He picked on the wrong kid. Johnnie Gravy was a good kid, according to his father, the town's judge. Sam would not leave Michea to defend himself. He quickly closed up the drug store leaving his wife Sara to bag-up the finished prescription orders. "Don't worry kitten," he told her. "We'll both be back in time for dinner!"

Nevertheless, it was now getting dark and Sara began to worry. Sam and Michea had been gone for hours! Why, even the phone stopped ringing at the store. Things were quiet now and people were home for the evening. Sara went over to the magazine rack and thumbed through some of the magazines until she found one that suited her.

And then the phone rang. At first, Sara wondered who would be calling the store after hours? Then it dawned on her; it had to be Sam!

She caught it on the fourth ring, "Charles Drug and Sundries," she announced over the telephone receiver.

"Sara it's me," Sam replied. "I need you to go to the phone book and call a lawyer right away. Have him come down here to the jail."

"Oh! Oh! Oh!" Sara exclaimed excited. "Is everything all right?"

"Yes, Michea needs a lawyer and …"

"A lawyer?"

"Yes a lawyer. Have him come to the jail as quickly as you can!"

"Oh! Oh, well give me your phone number in case-"

"Sara!" Sam yelled impatiently interrupting his wife. "I don't have a phone number. I am only allowed one call!"

"Oh dear, oh dear," fretted Sara. "Will you be home soon?"

"Depends on what this lawyer can do … just make the call and go home. Call you when I can … gotta go!"

Time drags on slowly when you're in jail. Luckily at least the both of them were allowed to share the same cell. Michea felt he had somehow been responsible for all of this.

Sam didn't like the fact that Johnny Gravy waltzed right out of jail an hour or so after being locked up. He didn't like the fact that Michea would be held until he had seen Johnny's father, the judge on Monday. And now, Sam was booked for assault on a police officer. Sounds worse than it was, when all it came down to be, Sam pushing the detective Henry Watson against the wall for insulting him.

When the attorney was notified, he called down to the station to inform Sam, that he would see him and John Doe later.

"I am fed up with people around here calling Michea John Doe." Sam shouted.

"I'm sorry," his attorney told him. "But until we can gather more information on him, Michea is simply an alias."

"The court recognizes John Doe but not his real name … oh now I get it! And because the Judge's brat is involved, as the criminal in this incident, we're the ones that will take the fall for all of this," Sam was really getting heated up. "What is it? Because I told you people I was pressing charges … lotta good that'll do. This town is full of corrupt law enforcement!"

Michea sank to the floor, his face in his hands. He never felt this way before. He felt ashamed. Things didn't work the same in this world. It all seemed to work backwards. Bad men are released to the streets while good men like Sam are put in a cage!

Now that Sam, being a registered pharmacist could not legally allow his wife to run the pharmacy, he was forced to keep the store closed all day Sunday and of course probably Monday too. It would be Monday sometime before the two would go before the Judge.

However; late Saturday evening, the Judge bit into a piece of steak bone and cracked a tooth. An emergency visit to a dentist was arranged and relief from the pain was on the way. The dentist was quite convinced the old repair on the tooth had come away from a good portion of healthy tooth, thus leaving nothing he could do but yank the tooth. And that he did, cracking loose, a bit of jawbone in the process. The dentist knowing he did a great job of extracting the Judge's molar, considering the fact, he was quite drunk still, let the Judge know that he would be experiencing a bit of 'discomfort'. That is in a few hours after the Novocain wears off.

"Don't worry sir," said the dentist. "I have the emergency number to the drug store. They'll open up and fill this prescription for you!"

In the apartment upstairs of the Drug Store, lived the Charles'. Only now, Sara was the only one there. Sam being in jail couldn't make it down to fill the prescription. But Sara was polite. "Why yes," Sara explained. "Sam is personally indisposed and cannot fill the prescription until his lawyer … um, I mean until he gets back from his business trip."

"Oh boy," said the dentist. "… yes Mrs. Charles thank you …" The dentist sobered up a notch and turned to the Judge. "The pharmacist had an emergency in the family and left town for the weekend … so the drug store is closed." The dentist turned and began rifling through his cabinet. "I have some codeine pills and ah, some aspirin I can give you to help you through some of that pain … er, I suspect you'll have!"

The judge already began to feel his jaw throb. And the Novocain hadn't even begun to wear off! No telling how excruciating this pain is going to be. He clamped down on a mouthful of gauzelike material and began giving harsh muffled orders.

The dentist had heard a lot of different types of muffled speech, but never like this! The nearest 'other' drug store was seventy miles away in Scranton and with a blizzard set to blow in, that would be an impossible option not worth considering.

"Look," the dentist finally said, "I can pop a little more Novocain in there to help you get by the next four hours or so ..."

The judge was well aware of all this, including the butchering he'd just been put through. He was also aware that Sam the pharmacist was in his jail!

Now the judge was a proud man. But as he sat in that chair, he came to realize that he was only proud when he wasn't in lots of pain. He reached into his wallet and handed the dentist a card. "Oh!" the dentist exclaimed. "You want me to call this number?" The judge nodded his head and made a clear verbal indication that he wanted him to speak to the pharmacist and give him the prescription over the phone. The dentist shook his shoulders but dialed the number on the card. "Yes sir," the dentist said evenly and obviously confused. "May I speak to the pharmacist? No I'm not kidding. This is Doctor Stevens. You see, I have a prescription order for the pharmacist-" The doctor turned to face the judge, phone still in hand. "They said I had the wrong number and hung up!"

"Gimme the phone," barked the muffled voice of the judge. The doctor stretched the coils of the receiver cord to the limit, giving the judge the receiver before re-dialing the number. The call went through when the doctor heard the judge say, "This is Judge Gravy, put the damn pharmacist on the phone now!"

"Yes?" Sam asked.

The judge handed the phone back to the dentist. "This is Doctor Stevens. I have a prescription for you to fill."

"If this is an emergency, you'll have to get a hold of judge what's his name! Otherwise, you'll have to make a trip over to Scranton."

"It's a prescription for Judge Gravy and he's in pain so I need this filled just as soon as you can get to your store ... are you in jail?" The judge could hear Sam's voice but couldn't make out what he was telling the dentist. "Wow," exclaimed the dentist. "You've been jailed?"

"Gimme the phone you idiot!" the judge exclaimed. He was already feeling some sharp pain seeping through the slowing down of the Novocain. The dentist looked back at the judge with a frown, then reluctantly handed the judge the phone.

"Watson," the judge barked. "This is Judge Gravy. Get the druggist into a car and get him down to his store NOW!"

There was a pause on the other end. "This is Sam Charles, the 'ahem, Pharmacist."

"Tell Watson to take you to the drug store and pick up that prescription!"

Another pause. "Sorry," Sam said. "No can do!"

"Okay okay," the judge snapped. "So now you want to make a deal with me huh?"

"Oh no sir," Sam replied politely. "I would never want to do something illegal like that! I have a Hippocratic Oath I swore on as a professional. So as long as I have criminal charges pending on me, my license is automatically suspended. I surely don't want to go to prison. So please sir, don't ask me to do something that would put me in prison!"

The pain was becoming unbearable now and the dentist was approaching him with another syringe dripping with Novocain. The judge wasn't sure it was working. "You drunken fool," he yelled at the dentist. "Except for my jaw, my whole damn head and neck is numb!" The judge paused to quickly compose himself. "Mr Charles," he said nicely. "Please hand the phone to someone standing by."

Sam listened to Watson taking orders from the judge. He then hung up and scrambled about his desk getting some papers together and yelling at someone to come up and ride with him to the drug store. "Mr Charles," Watson said. "You have been exonerated of all charges. We are to meet the honorable Judge Gravy at your drug store where he will sign all the paperwork to make it happen."

"Great," said Sam. "Come on Michea, we're going home!"

"Ah, wait a minute," Watson said sternly, "just you sir!"

Watson opened the cell door and Sam just stood there. "Nope," he said to Watson. "Ain't going nowhere without Michea."

"You have to. You've been released!"

Sam smiled slyly, "Where's the release forms? Are they signed?"

Watson, well aware of the situation had no time to argue the point. "Here," he told Sam handing him his folder of paperwork. "Let me get John Doe's and we'll all go together and be released from jail and I will be released from my duty and my job … yada yada!" And Henry Watson did just exactly that. They all climbed into a police car and headed down to the drug store.

Moments later all parties met at the door of the drug store as Sam rattled around with his keys. Once opened, Sam punched in the alarm code and disarmed the security alarm system. Behind him followed Michea, Watson, the court notary, the judge and his drunken dentist. Sam showed his hospitality by having them all line up at the soda fountain bar and take a seat. The paperwork was spread out and the judge began scribbling his signature on page after page of documents. It took a little bit of time. Time that increased the pain on the judge, but he was relieved in the thought, he would have his pain killers. Michea took the liberty to fix everyone an ice cream sundae!

It never really made sense to Michea, what all this paper signing was about. What did become clear to him was the urgency to find a way back to his time. In Michea's mind were only two approaches to solving this riddle. He would find a Medicine Doctor, who could guide him, or he would go and seek answers from the Turk Street gang. They were a tough band of hoodlums who hung out all night long in the Alfred Brown Elementary School playground. Michea saw a group of these men visit the store. Along with this street gang, John Gravy was seen being loud and smoking cigarettes.

Michea watched and listened whenever he had the opportunity. John had a distinctive voice. A voice that was reminiscent of the voice of the punk who tried to rob him the moment he fell into this time period. Approaching this punk could result in getting Sam in trouble and Michea couldn't allow this to happen to his best friend. If only he knew for sure this punk had his spear point, he would take it from him, whatever the cost.

It was late and very quiet. He was certain Sam and Sara were in bed and asleep upstairs. Michea put on a warm jacket and boots. Moving to the backroom security door, he reset the burglar alarm the way Sam showed him and then, slipped out the door into the night.

Chapter 14:

The Turks

The neighborhood landscape was thickly blanketed with the heavy layers of freshly fallen snow. The blizzard from a few nights ago had dropped ten inches of snow and now sidewalks and streets were lined with higher walls of plowed snow. The crunching of snow under Michea's boots made him wonder how anyone could hunt under these harsh conditions.

Michea stopped. In the air carried a mingling of voices. Sounds that skipped over the snow in the crisp evening air told him he would need to make a turn at the next street corner. Halfway down the block he heard the slam of a car door and a motor rumbles to life. Then the motor roared and died down, a few more times. Suddenly, the motor roared back to life and a loud scream of tires pierced the cool air as the car swept past Michea. The headlights were not on and the car spun sideways, one way, then the other; while, down the street, came a burst of laughter and whistles. Michea could make out three; maybe four dark figures clapping their hands over the light of a trash can fire. There, in a bus shelter they stood passing the time loudly exchanging stories. Michea quietly made his way past these gang members. Michea lurked from shadow to shadow knowing the fire blinding the eyes of these men made for his advantage. Before long, Michea stepped away from behind the bus shelter, taking his place beside them at the fire barrel.

"Holy Shit!" exclaimed the tallest of the four standing there. "Who the hell are you?"

The other three frantically looked around them as if they feared retaliation from members of a rival street gang.

"You alone," asked the one standing next to the tall one. He had a stern look in his eyes. His mouth pulled down to the deepest frown Michea had ever seen. "So what the hell do you want, Indian man?"

"Nothing from you," Michea said evenly. Across the barrel from him stood John Gravy; Michea pointed at him saying, "He has what I come for."

Gravy turned a look to his friends. "He's the sap that got the drop on me at the drug store. I think we ought to teach him a lesson!"

Michea didn't appear moved by this threat. He just stood staring at Gravy.

"Did you hear what I said?" John Gravy's voice raised another octave.

Nobody, not even the other three gang members moved to take issue with Michea's request.

Finally, Michea said. "And so, what is your lesson?"

The largest of the Turks busted out laughing. Raising a finger, he pointed to Michea. "Man, it's like this dude is enjoying this!"

Michea for the first time, smiled. "Yes," he said. "I enjoy this!"

One of the others looked edgy. His eyes shifted between Michea and the others. "Man, I don't like this ... this dude probably has help or a gun or something!"

"Aren't you guys gonna do something," screeched John. "He's only one guy and we've got him outnumbered!"

"Shut the fuck up John!" said the gang member with the deep frown and the seriously stern eyes. "If you have something of his, you better give it back!"

"Are you serious?" John screeched.

There was something about Michea these Turks had never seen before. Something in the way he handled himself. And what they read in his eyes, they had never seen before. It was primeval and it scared them. Conversely, Michea was not in the least bit frightened of them.

"What do you want?" John finally asked Michea.

Michea waited for a moment. He just stared into John's eyes looking forward to seeing a hiding of guilt. And there he might have seen a tiny spark of it, but he wasn't completely sure. "The spear point you took from me when I was down," he asked finally. "Where is it?"

Michea saw John slide his hands across the pocket of his leather jacket. It was a very slight move, but it was a move; spawned perhaps, out of guilt.

"I don't know what you are talking about," John replied. His eyes shifted as if he was startled about something, or bewildered for a moment. "Are you calling me a thief?"

"And what if he was?" one of John's buddies snarled. "What are you going to do about it?"

John Gravy had met his match and now he stood alone, just him and Michea. "So are you guys gonna back down on me?" he cried.

"No John," said his friend who's deep frown turn right side up. Smiling now he continued, "Why don't you dump your jacket pockets and show us what you got?"

John Gravy huffed. "I don't have his damn spear head." Michea watched, hoping John would finally produce for him his ticket out of this time. However, all that would come was a pack of cigarettes, a few sticks of gum and a brand-new switchblade knife. Michea's eyes flashed at the sight of that knife. He'd never seen such an intriguing knife that hid away like a panther's claws. He wanted to have such a knife, but the police took away the one left behind at Sam's drug store. Being the wise guy John Gravy was, he noticed the expression of interest in his new knife. "Ya like this baby?" John asked, while flashing the pearl handled switchblade in Michea's face.

Michea wouldn't give John the satisfaction of bartering with him. There was no place in his tribe for a knife such as this anyway. "You will give me the the spear point." he said flatly. "Then maybe, I will barter for your knife."

The lucky obsidian arrowhead was in Michea's clenched hand. He wanted to show these guys what his spear point would look like. By showing them this obsidian arrowhead he could better describe it by telling them it was twice bigger in size. But then it seemed the meeting was about to come to an abrupt halt. Barreling up the street came the very same car that had sped away earlier.

The car nearly slid to a stop next to everyone standing by the fire. The driver ducked his head out of the window and yelled, "Guys! Tony swung a deal on 9th and Main. He's buying the pizza!"

Everyone jumped into the car except John who now took a fighting stance in front of Michea. "Come on Gravy," yelled one of the hoodlums. "Let's go!"

Seems John wanted to take a stab at Michea before making his getaway. It happened very fast. John clicked his switchblade open and said something like, he was going to break his new knife on this Indian.

He lunged forward. Michea lifted his foot and landed it on the side of the knee of John Gravy. A sickening snap was heard above the rough and loud idle of the automobile. Michea stooped down and picked up Gravy's switchblade knife. He studied it while everyone piled out of the car to get a look at John's twisted leg. "He broke his fuggin leg," exclaimed one of the hoodlums. John was screaming and yelling to be taken to the hospital.

Michea was trying to figure out how to close the knife, but it was locked open. "Ah," Michea said smiling, "there is a button!" With that he closed the blade back into the handle with an audible click.

Michea knew they might attack him, so he stood waiting. "Back off," said one hoodlum pointing to Michea. "He's got a knife too!"

Taking one last look at the knife, he tossed it through the open back door of the car where they had dragged John. "Not worth more than two dollars," Michea said, crossing his arms.

"Your ass is grass," yelled the driver just as he punched the accelerator on his beat up 55 Chevy.

He knew this yell was one of a retaliatory threat. But the expression did nothing but amuse him. He smiled and waved a goodbye. He repeated the phrase over and over, shaking his head in amusement. "Your ass is grass ..." *When did the world come to this*, Michea wondered. Maybe it meant fear in one's heart to hear this. It did nothing for Michea, except make him laugh.

Quietly, he slipped back through the alleyway to the rear of the drug store. There would be lots to do tomorrow Michea knew. The Spear point was in this world's time frame. But where, Michea could only wonder. And there were lots to wonder about now. *What was the purpose of John's attack on him? Was he hiding something? And if he did have the spear point, what good would it serve him? Especially in this world where there are such wonderfully crafted knives and weapons. Did he know of the Spear-point's power?*

Michea lowered himself to bed. In his mind, he tried to sort out all the possibilities connected to the missing spear point. Michea tried to think back on the origin of the spear head. It now became obvious the spear point's origin began with Huega's discovery of the dark rock from which it was made; his Mother, being the one who crafted this enigma. It was himself who brought this spear point to this time. However, taking on such thoughts for their answers were a tough proposition. It would be something that only a learned man could only answer. So the subsequent step, Michea decided, was to find such a wise man who could understand these things. Perhaps even guide him to the next step of how to get back before the settlement is attacked!

Chapter 15:

Threatened

Sam was up and out the front door of his store before the sun had a chance to shine its light on the town. With a scraper and a bag of salt in hand, he was scraping. Around the back-alley corner of the store came Michea with his scraper and bag of salt.

"Ah," said Michea with a smile and pointing at the sidewalk. "Mrs Smith's footprints!"

Sam laughed. "Well Michea," he said cheerfully. "You're an Indian; you tell me how old are Mrs Smith's footprints?"

Michea knelt down and swept some loose snow from one footprint, then said, "One, maybe two hours old."

"That would have made it between five and six 'o clock!"

"Mrs Smith," Michea added. "has an injured left ankle."

"You could tell that?"

"Yes," Michea said, stopping to rub his nose. The cold made his nose run and itch sometimes. It got cold in Florida, but never so cold as to make one's nose do this.

"Well," Sam began. "You'd be happy to know Mrs Smith has gout and I fill her prescription here once a month." Sam shook his head as he was amazed at Michea's footprint assessment. "And yes, she favors that foot!"

Michea smiled. "Now it is my turn!" he said enthusiastically.

Sam looked over to an eager young man pointing to the side of the drugstore wall. There to Sam's surprise was a junket of black spray paint scribbling that spelled out a deadly threat to Michea's life. "What does that message mean?" Michea asked.

"It means we need to finish up here," Sam began nervously, "and get inside!"

Michea knew something was up when the police arrived to take a look at the wall. Sam was on the phone to the hardware store for a can of white paint to match the wall and a paint brush. Sam didn't want Michea out painting a wall, since that makes Michea a perfect target. So instead he sent Michea a few doors down to the hardware store to pick up the paint and brush. This too turned out to be a bad idea when Michea was hit by an old 55 Chevy while crossing the street corner near the drug store. Sam ran out of the drug store when he heard the squeal of tires burning rubber and the thud of Michea rolling over the passenger side fender onto the sidewalk. Michea landed on his feet, never dropping the paint and brush he carried!

"Your ass is grass!" Michea yelled at the car as it sped away.

"What the heck!" Sam yelled shaking his head. "That kid tried to kill you!"

"Yes," Michea said as he approached Sam and the doorway of the store. "I never have my apron on when I need it!"

Sam's heart was pounding. Obviously Sam grew fond of Michea and he was afraid for him. "Now we learn to paint!" Michea said holding up the can and brush in front of Sam.

"Well," replied Sam. "It's really way too cold to paint … that is, unless you are really really good at it. Takes experience to know how to paint in the winter!"

"It is a good time to learn!"

Sam finally gave in saying, "Yeah why not. The worst thing that can happen is they slam a car into the building, trapping us both like bugs!"

The rest of the day was business as usual, no further threats or attacks. Still, Sam wasn't himself. He seemed a bit worried because he knew sooner or later someone could walk into his place and take aim on poor Michea. He had to find him an alternative place to stay. Maybe even disappear for a while, until as they say, the heat cools down!

"What," Sam asked.

Sara stood with her hands on her hips. "You've been daydreaming a lot Sam. You will never get that hose thingy finished. I was talking about the brownies. When they cool, you and Michea are welcome to have a few."

Sam smiled up at Sara. He was putting on a new supply line to the sink faucet at the soda bar. "Yes Dear."

Michea was already there, putting a few brownies aside for Sam and himself. The rest of the pan, Michea placed in the display case near one end of the soda bar.

A minute or two later, Sam was wiping his hands off in a dishtowel, "What ya say we call it a day."

Michea shrugged his shoulders. He knew Sam well enough now to recognize a plan brewing in his head. Sam mentioned earlier this day that he'd like to pay a friend a visit. Michea was eager to go with him. This friend of Sam's sounded interesting. He owned a gun store. And Sam promised Michea that he'd find Glen to be quite interesting.

Sam timidly smiled as he lead the way out the front door of the drug store. "Well," Sam began. "Glen's place is just a few blocks down. With luck we won't get run over by a car!"

The front window of Glen's Gun Shop had a full sized Deer with a sign on it that read: Only 25 dollars. Michea stopped to stare at this phony deer. "This one, not feed too many!" he laughed.

"Come on," Sam said holding the door. "It's just a practice target!"

Now here was a place after Michea's own heart! Every step he took was measured with equal attention to all the fascinating items inside this place. There were real deer heads mounted on the wall and all sorts of taxidermy samples at every turn. The voices of Sam and his friend Glen just melted away, in the distance.

"Been awhile Sam," Glen said. "Stopped by the other day and Sara said you were away on business."

"That's sort of the reason why I am here Glen."

"Something to do with the cops and fire trucks you've been having around your place lately?"

Sam nodded his head and called Michea over for Glen to meet him. "The Turks," Sam began, "have been collecting on me for a long time!"

Glen looked around himself. Surrounded by firearms he replied with a grin, "Never had that problem here!"

Michea stared at Glen, interested in the way he carried himself. Very smart man Michea knew. Finally, all conversations stopped as Glen stared back at Michea. "So you stood up to the Turks and nearly put one of them in the cemetery Sam tells me."

Sam excitedly butt in, "Like I said Glen. If I hadn't stopped him he would have!"

Michea never challenged Sam's rendition of the incident. He had other subjects of thoughts running through his mind. He kept his attention on Glen. "Are you a Doctor?"

Glen's eyebrows rose. "Strange you should ask," he said. "Two years of veterinary school and my record showed I was lousy at saving the animals, so I decided to stuff the damn things instead!" Glen stood up from his chair and with outstretched arms pointed to all the wonderful animal specimens he stuffed. "Taxidermy … you can't beat the money nor the charm!"

"You bet," Michea replied. "I see you are Indian."

"Yes I am. Full blooded Apache," Glen said proudly. "And you?"

"Ponctoan."

There was a long pause. They stared at one another. A strange kind of stare, Glen's eyes narrowed. "You know I am an historian?"

Michea shrugged, "but not a Doctor."

Glen sighed. "You know, I have had the privilege to have met Dwight D. Eisenhower and the misfortune to have met Fidel Castro but never ever a Ponctoan!"

"And which is your fortune today?" Michea asked, drawing a look of shock on Glen's face.

Glen thought a moment. Sam was keeping out of this. There was something quite controversial about this conversation.

"The Ponctoan," Glen began. "Their homeland … was?"

It was hard for Michea to follow conversation here. There were so much interesting things to look at here in Glen's store. He heard Glen's question and merely replied, " Mayaimi."

It appeared Glen expected this young man to reply with a State like Florida, as that is where the Ponctoan was said to have lived. That is, if this young man were just pretending to be Ponctoan and had read up on the tribe. But Mayaimi? Glen mouthed the word silently and then held his chin thoughtful. "Michea, are you saying your settlement was on the Big Lake?"

Michea nodded and smiled his head. "You bet."

Glen shook his head. He seemed to be confused. "Man, you got me there! Can you tell me where your people went?"

Michea pointed back toward the drug store and said nonchalantly, "They are, in another time."

Glen was certain this young man was an Indian. His mannerisms and even the way he walked was true disciplined Indian. A warrior type. He walked with a perfect walk. His feet; toes lined straight forward so that when his walk slowed his feet went heal to toe. Glen knew this man was a 'silent stalker' of animal and human beings. And although the response Michea gave of the whereabouts of his people, could not be disputed, Glen became drawn to him. "I have something interesting to show you," he told Michea.

Over in the corner of the store, a dusty crate and few paper sacks were kept behind the counter. Glen carefully lifted these items up and placed them on the counter in front of Michea. Out of the bag, Glen poured forth a copious amount of Indian arrowheads. Many of which Michea recognized as enemy tips. Mostly, Calusa arrow tips made of shell. "These are all Florida Indian weapon artefacts." Glen announced. Then carefully Glen pulled the lid back on a small wooden crate. He reached in and pulled out a fistful of wood shavings. With his other hand he fished out a large obsidian weapon tip.

Michea gasped. His eye wide, he lowered his hand slowly over the piece. He knew this artefact for he had used it only weeks ago! It was his favorite sabre tip! The very tip was missing, broke off perhaps in battle, a long time ago, Michea surmised. Michea picked it up. It was his Mother's work in every detail! His eyes shifted up to look at Glen looking at him. "You have a very special battle tool," Michea said reverently.

"It appears to be a long spear shaped knife," Glen explained. "It was purposely curved for reasons not quite clear."

"It made enemy die fast," Michea said. "Curve made head come off fast!" He made a slow swing with the obsidian sabres to demonstrate the effectiveness of his words. "Enemy go down and not come back up!" Michea picked up a pencil from the counter and the receipt book. On the back of a receipt he drew the stone sabre, bound to a cypress pole about six feet long. He showed the drawing to Glen.

Glen marvelled at what he saw. "I see clearly now what you mean about the effectiveness of such a weapon!"

"So there is a good three or four inches missing off the tip," said Glen. When Michea asked where he found the sabre, Glen told him he bought it from an artefacts vendor at a gun and knife swap meet. "The man said it was dug up by a sugar cane farmer in South Florida. Thought there might be a chance it came from the Ponctoan Indian People!"

Michea nodded his head. "It is Ponctoan."

"Man, you say that with such authority!"

"That is because," Michea said with a sullen expression. "It served me well in battle." Michea set the sabre down on the counter for the last time and just stared at it. "I used it also to bleed-out man and pig ... like the Warrior Prince to the Calusa Tribe."

Sam smiled. "You had to ask," he said to Glen sarcastically.

Glen crossed his lips with his finger, calling for a moment of silence. He saw a tear run down Michea's cheek.

Unknown to Glen and Sam, the memory of that night brought the image of his newly wedded wife Lichea, in his arms. He drew a deep sigh and a sniff. He had enough of this talk and thanked Glen for showing him his collection. Michea saw the pride Glen had for his collection. He was having a special display case made for all his Indian artefacts. So Michea wanted to donate more. Reaching into his pouch, he pulled out a half dozen arrowheads, some of which his Mother made him. His favorite, the big one was his lucky one and only showed it to Glen before placing it back into his pouch for safe keeping.

Glen's eyes squinted in the detail of one obsidian arrowhead. It was indeed the same style of that which the sabre was fashioned. Glen was lost in history for a moment it seemed when Sam finally broke him out of his trance. "So we're here to take a look at buying a pistol."

Glen never looked up from his arrowheads. "Really," he said, "what sort of handgun are you interested in?"

"Actually, a gun that can easily be concealed. Like one of those that uses a clip instead of a revolver type, I guess."

While Glen was guiding Sam's purchase toward a handgun, his eyes strayed to see what interested Michea. Not surprising to Glen, Michea found the isle where all bow and arrow supplies were displayed. Being full-blooded Apache, Glen felt himself spiritually attached to the bow and arrow. Glen was an avid Bow Hunter that took him to all parts of the country on hunts and tournaments alike. So it wasn't surprising to Sam that Glen had a shooting range in the back of his store. Only a small section reserved for firearms testing, with the largest portion for bow and arrow testing.

It wasn't long before Sam felt he had been pushed to the back seat of this discussion. Glen was gathering up an armful of arrows and following close behind was Michea with four rather nice looking bows. With everything laid out on the demonstration table, Glen began a discussion. "Tell me what you know about these bows!"

Michea was the proverbial *'one eyed cat in a seafood store'!* He was excited; for he had never seen such fine weapons! Michea had a hundred questions, but did not know how to ask a single one. The only bow that even resembled the bow he was used to, he left behind on the display rack.

"Well," Glen said, reaching to pick up one of four bows Michea chose. "This one is pretty close to what you should be familiar with. It is a 'longbow' and unlike the old ones of ancient times, this is constructed of laminated layers of fiberglass."

Studying it up close, Michea rubbed the long rib of the bow between his fingers. Slowly he studied the string attachment and the lay-up of the rubberized grip which fit his hand like as if it were an old reliable weapon made just for him! Michea looked over to Glen, who was holding out to him an arrow so elegantly crafted that even the feathers were sculpted with a style never seen before. Michea set the bow down and accepted the arrow. He wanted to inspect this arrow in all its detail. It was as light as air! He held it up to one eye. It was keenly as straight as any arrow he had ever seen.

Glen held out his hand for the arrow Michea was studying. Michea graciously relinquished and Glen promptly unscrewed the arrow tip and then the notched end including the feathers! Laying them back on the table, Glen said, "Go ahead. Pick up the pieces and put it together."

Slowly Michea obliged Glen's request. He studied the small ridges on the arrow tip. These were common threads and Michea had learned all about threads! Many bottles and jars in the drug store had them, but he never imagined an arrow having threaded parts! Michea let out a big laugh. He was truly impressed. He looked at Glen, "You are a great arrow maker!"

"Well thank you! But I prefer to buy them from an outfit in Michigan." Glen pointed out to the bull's eye target, a distance the length of a bowling lane away. "Go ahead; test your skills on that target down there."

The bull's eye target was an object whose rules transgressed time. Many of Michea's days were tested by his archery skills. He practiced those skills by setting up several hoops made of soft pliable willow wood. Hanging hoops from short tree limbs where a perfect pass through would allow the arrow to hit center of the tree, was quite a challenge that paid off for Michea. Not only that, Michea became highly skilled at hitting the mark on moving objects such as squirrel and even flying birds.

Michea nocked his arrow and drew back the bow string. Slowly Michea released his pull as if studying this new type of long bow. The characteristics and feel were very different. *Very smooth*, Michea thought. Again he did the same and then again, this time lowering his bow to take aim on his target. Within a microsecond it seemed, Michea let loose his arrow. The arrow found its mark, dead center. "I like this bow," Michea said.

"Wow! You should," Glen replied. Pointing to the next bow on the table, "You might want to check this one out." Michea set the long bow down and picked up the recommended bow. "This one is what is known as a Flat Bow. Instead of the rounded arms, you will notice they are flat and very wide. Try it, you will find it quite unique."

Michea immediately began stretching the bow string, interestingly questioning the integrity of such a bow with flat arms. The question was on his lips, but he never asked it. Instead he picked up an arrow, nocked it and drew back on the target. Michea never heard such a sound. This bow really packed a punch and he could feel the back flow of the power in his outstretched forearm. He put the arrow right alongside of the first arrow on the target.

Sam and Glen looked at each other, amazed at Michea's archery skills. "He's an Indian alright," said Sam sardonically.

"This one," Michea said holding out the Flat Bow, "even better!"

"Whoa. Whoa," Glen said slyly, "try this one." Glen held up a bow with a different kind of curve in it. "It's called a Recurved Bow."

This was heaven to Michea. So many different kinds of Bow Technology, it baffled him. The Recurved Bow had an interesting way it was strung. Almost as if it was strung backwards! Michea tested the pull on this bow. It had a robust tension that increased evenly as the string was pulled. Slowly Michea returned the string to rest. Selecting a brightly colored arrow of orange and yellow, he nocked, pulled and released the arrow with a snap. The arrow hit its mark precisely alongside of the second arrow. Then, just as quickly as he released the arrow, Michea flung the bow over his head letting it come to rest with the bow string across his chest. The Recurved Bow design allowed the bow to rest secure against his back where his quiver of arrows would also be. "This one," Michea said. "Good for war and battle from atop trees."

"Who could argue that," Sam said flatly.

"Yeah, he has a point," Glen agreed.

The Recurved Bow seemed to be the bow of choice but then Michea took an interest in the last remaining bow on the table. He twisted his head sideways as if trying to figure it out. It was the ugliest bow he ever laid eyes on. It was lightweight and smaller than all other bows he had just sampled. Further examination of the bow, revealed a material unfamiliar to Michea. And the way it was strung was equally confusing. Glen saw the look on Michea's face as he slowly examined the bow from end to end. "Yeah," Glen said with a sigh. "It's new and basically a prototype bow. Apparently, the inventor intends to develop a market for this Frankenstein of a bow. As you can see," Glen said pointing to the engraved message on the bow. "Patent Applied For."

"What's with all the pulleys?" Sam asked.

Glen began by telling Sam and Michea that it was an example of the science of applied physics. The way this bow was strung, it had to run through a pulley at one end before it could be brought out to the drawstring. And because it was a bilateral device, the string had to go through the same pulley process on the other end. Each pulley allowed a 2 to 1 purchase over the force it took to pull back the drawstring. Total in all, a reduction of force to pull the arrow back would be 4 to 1. One part input energy to an equated 4 times greater output.

Michea pulled back the drawstring on this strange device of a bow only to find the pull suddenly give way to a feather touch on the string. Michea looked to the bow's arms. He was certain he had broken one or both! He carefully studied the action of the bow while he pulled back on the string. The arms were still there; and with a fiercely arched posture that reminded him of the powerful black hind legs of an adult panther! This Michea knew would bring 'arrow-ready' to a new dimension. He no doubt could hold a fire ready arrow for long periods of time! He slowly released the tension until he felt the nudge between where the pull of the string went less to that of more. He quickly pulled up an arrow and nocked it in place. The shaft came to rest on a specially designed groove which further impressed Michea. Pulling back to the spot where the increased pull suddenly broke free, to an easy to hold spot. He also noticed it was placing the arrow precisely balanced in overall length on the bow itself.

"It's what the inventor calls a compound bow," Glen stated flatly, "haven't had the chance to try it out yet. Came in just yesterday. I'll be writing a magazine review on it in a few days."

When Michea let the arrow sing, it sang right straight through the bull's eye target and into the thick plywood wall a few feet behind it.

Michea reverently lowered the bow to cradle in his arms. "It is magic," he whispered.

"Well, there's your review," Sam said dryly.

"Wow!" Glenn exclaimed. "You know, I have to run tests on that thing. Such things as ballistics, like how many feet per second that sucker can spit an arrow and so on." Glen paused to watch this Indian stroking this bow like a lost kitten. "Of course I get to keep it for my troubles … writing the mag review and all. And since he gave me some mighty fine arrowheads, the bow is his free of charge. In a few days come by and pick it up!"

Slowly Michea set the bow back on the table. "It is the best bow," he said with certainty in his voice. "It would be every mans' dream to own such a bow as this one!"

"This time next week Michea," said Glen. "stop by and pick her up!"

"You bet!"

"Say Glen," Sam asked. "You got anything that shoots lead?"

Glen looked at Sam with a grin. "Step right this way young fellah, to the 'plugged nickel shooting arcade'! Was just testing a snub nosed .38."

"Isn't that a revolver?" Sam asked. "I thought I made it clear, I want a semi-automatic pistol. One that uses clips."

Glen rolled his eyes back. "Trust in your local gun smithy Sam. Say let me introduce you to speed loaders!"

"Don't bullshit me Glen. I'm a pharmacist. Ya never want to bullshit a pharmacist!"

"Well, ya know Sam. Come to think of it, I do have a nice little .32 semi-automatic over here. Fits in your pocket like a pack of concrete cigarettes!"

Sam smiled. "Perfect."

Michea stood by watching and listening to the tinkling of shells and bullets, or bullets and shells. He didn't know which was which. Sam seemed to know a lot more of this gun thing. And of course Glen was a real expert at this too. Glen was an all-around expert on weapons! But after firing the guns at targets and after all the bang noise, Michea cared not for guns, mostly because they didn't smoke like the one that shot poor Lichea. Those were terrifying guns! Guns crafted for the elite warrior of the white skinned society. Those of whose tactics included nimble footed animals of loyalty called horses. Michea recalled being chased by one such horse and a man with a sword. Luckily, those types of animals know nothing of climbing trees!

But then there are some who know nothing of pursuing a battle strategy. So it was with one of the Turk's gang members. He was an entire head taller than Sam and Michea. And he appeared from out of an alleyway with the expressed idea of blocking the sidewalk. Sam was familiar with this tactic. It was called 'bullying' and Sam told Michea that they were about to be 'bullied'.

"The bigger they are," Sam told Michea. "the harder they fall!"

Michea, of course, never heard such a wise saying as this. Judging from the size of this guy, it would have to be true. But this big man was not alone. The stage was set across the street. It was in this ploy, that Sam and Michea would dodge this big guy and cross-over to the other side of the street. Or so it was planned by the Turks.

It was growing dark and in the alley across the street, Michea spied a cigarette glow in the shadows. It was becoming clear to Michea that the strength of the Turks was in their numbers. Singled out, they were harmless as bear cubs.

Sam, not aware of the potential ambush that was assembled in the alley across the street, looked for a break in traffic. Michea grabbed Sam's arm. "No," he said flatly. "We send this big one on to his animal side ..."

Slowing pace, Michea slowly reached into his pouch and produced a razor sharp shard of obsidian. Michea smiled, "... he looks to become a woodchuck."

"Should of bought that gun," Sam snarled under his breath. "I know he has one!"

The big man motioned with his hand on the breast of his leather jacket that he had a gun.

"Now Michea," Sam ordered, "we're defending ourselves. Let's not kill him!"

Michea understood the order to take this man back to the store alive!

Michea and Sam walked right up to the big guy. Sam said a few things while the big guy grunted a few threatening slurs. Michea was busy disrobing from his jacket and yanking off his shirt. He swiftly slashed a few long strips of material from his tee-shirt.

The big man's attention was quickly attracted to Michea and then, so was Sam's.

"What the hell is he doing?" The big guy asked Sam.

"Don't ask me," Sam replied just as puzzled.

"What's up man?"

So now Michea was known here only by 'man'." It irritated him for no other reason than he sensed he didn't like this person. "I am making you some much needed bandages ... woodchuck!"

Sam was taken aback by Michea's arrogantly composed ploy. "That's right Chuck," snickered Sam. "Keep your hands visible! You're gonna walk with us a block down to the tavern …"

"Fuck you," he replied reaching a hand into the lapel of his leather jacket.

Michea swiftly gashed the entire back of the big man's hand, effectively cutting tendons that operated his fingers. This was move number one in Michea's mind. Second move was to send him off to the animal side. But Michea would wait to see if step two was needed. Nervously Michea kept an open eye about him, to the alley across the street and behind them as well. Sam's eyes agape with the sight of this big guy's hand gushing blood like a ruptured hose. The big guy slowly withdrew his hand from his lapel and with a puzzled look on his face began to plead with Michea not to kill him.

Michea nodded once and then looked to Sam saying, "Fix his hand with a bandage and then we take him to the drug store for medicine." Sam went to work on a big man whose knees were shaky and giving out on him.

"Take his gun," Sam ordered Michea, looking back from the big man who had just slipped to his knees. "You're gonna be fine, now get up and start walking!"

Michea took the gun and then threw it into the darkness of the alleyway where the big guy had previously hid. "Great," Sam muttered under his breath. "I was prepared to pay seventy bucks for a gun and the kid throws a good one away!"

Two blocks Sam and Michea staggered under the weight of this big guy. As human crutches, Sam's back was complaining and he let it be known. The rest of the Turks Gang that was laying in ambush scurried away in the night like the rats they were. They weren't about to be part of the cops and ambulance scene. But then when they reached the store and the big guy was safely seated at the soda bar; he pleaded with Sam not to call the police. "You're gonna need stitches, you big dope," Sam snarled.

"But the cops," he pleaded. "I gotta record and something like this will put me in jail!"

Sam had a soft spot in his heart for mankind and he could see this guy was just a loser looking for a purpose in life. "Look Ray," Sam began, after becoming more acquainted with the big guy. "You are going to need surgery on that hand. I'm going to call an ambulance and they will take you to the emergency room." Ray shook his head agreeable that Sam was looking out for his best interests. "Again," Sam told Ray, "Your story is that you cut your hand on some sharp metal, climbing over a fence … I mean that's what you hoodlums do anyway right?"

"Oh yeah," Ray replied, "jumped a fence just a few days ago behind the bakery down the street!"

A surprised expression crossed Sam's face. "You broke into a donut shop and risked going to prison, for a couple of donuts?" Sam began to wonder, what other kind of information he could extract from this stooge. "Well," Sam followed up saying, "I wouldn't use *that* fence as an example … that is, in case anyone asks."

"Right," Ray nodded knowingly.

"You were running from a dog and jumped the fence behind the drug store here okay?" Pausing, Sam continued, "Those guys you hang out with … "

"The Turks?"

"Yes," Sam said. "Tell me about them. Why are they after me?"

"Oh come-on man," Ray mused. "You know I can't talk about that!"

"Oh really!

Sam disappeared behind the pharmacist counter and quickly returned with a diabetic syringe filled with tap water. He approached Ray with the syringe, the needle glistening with droplets of water oozing out from the tip. "Tell me what you know Ray," Sam said. "Or I will get it out of you medically!"

Ray's eyes drew large as saucers. "I hate needles man!" Ray swallowed hard. "That ain't that sodium penthouse stuff is it?"

"Sodium Pentothal my boy," Sam snickered. "Yes the truth serum. You are about to tell me anything I ask."

"When's that ambulance coming?" Ray asked nervously.

"Just as soon as I get my answers I will make the call."

Sam got his answers and called the ambulance all without having to make further threats to Ray. Ray told him everything Sam wanted to know, everything from why they wanted Michea gone, which was obvious. Collecting insurance money from the drug store was difficult with Michea there. And Sam was proud of that. Nevertheless, when it came to the Turks plans to jump and kill Michea, Sam knew he needed to send Michea on his way.

With all the excitement over now, Sam sat alone pondering what to do with Michea. In deep thought and the drug store past hours, the door remained unlocked. Sam suddenly looked up to a noise at the door. Someone was coming through the door and Sam gasped at the thought of being visited by street hoodlums. "Who's there?"

There was no answer. Instead, Michea appeared smiling. He had a surprise for Sam, "Here is your gun!"

Sam blew a sigh of relief. "Are you crazy?" Sam scolded Michea. "You went out in the night and into a dark alley to get that gun?"

"I heard you say …"

"I know. I know," Sam confessed. "I just didn't think you would know what I said."

Michea just smiled, "I understand more than I can say." Sam studied the gun. It was exactly the gun he was looking for. "I also know it is time for me to go … and you will need such of a gun to protect you and Lady Sara."

The gleam left Sam's eyes and he swayed his head back and forth a few times. True, it was time for Michea to go. But with Michea promising to stay the night, Sam in turn promised he would find Michea a new place to stay.

*****

The only consolation Sam got out of dropping Michea off at the Boys' Home was that it was in a better part of town. The college brought the neighborhood up a good notch or two and as far as Sam remembered of the area, it was scoundrel free! Before Mac past away, he had renovated the place and added five more bedrooms to the back of the Boys' Home. As long as there was a spare bed, Sam knew he wouldn't have to beg … much.

When Maggie Charles answered the door, Michea walked passed her like he owned the place. It startled Maggie and her hand came up to rub her throat. "Oh my," she began nervously watching the dark stranger brush past her, "is that you Sam?"

With her attention now on Sam who stood outside the door in front of her. "Of course it's me Sam." For everyone there is a personality that they are known for. Sam wanted to try to maintain that personality. It made him feel he had the 'upper-hand' on most situations. But Sam sighed when Michea just walked into the house. There just wasn't anything left to surprise Sam after this. "You'll have to excuse him, for blowing my ego Maggie … "

Maggie's brown eyes softened and she smiled. "Sam, what a surprise!" She reached out and gave him a big hug. "Well, come in," she begged. "It's freezing out there and I have five gallons of chili simmering on the stove!"

"Oh, my favorite," Sam replied, stepping into the foyer.

"It's Mac's favorite too!" Maggie said enthusiastically, but then hesitated in thought. Her face drew a sad overtone.

It had been going on two years since Sam's brother Mac passed away. But it wasn't Sam's way to dwell on sad occasions. "About that chili?"

"Yes?"

"Ya think it's safe to feed an American Indian chili?"

Maggie looked a bit confused. "Do you think he's in the kitchen?"

Sam saw a sparkle in Maggie's eye. "Well," Sam began, "if he isn't, he oughta be!"

"You are so right Sam," she said. "Let's go to the kitchen and have some!"

*****

Time slipped by quickly. Already Sam had a half hour left to catch the last bus home. He had talked highly of Michea and explained all the trouble he had been in, but assured her he was a good young man and that he would be happy to pull his own weight around the home. Sam saw Maggie was finding it 'somewhat' difficult to connect or bond with Michea.

"Well you know," she said, "I don't have any babies here anymore Sam. All adopted. And the ones that weren't, have grown out of diapers, so life is easy here now."

Confused, Sam needed a solid yes or no as to whether she would accept Michea here at the home. "He's a fine young man. He'd be happy to do all the chores you can lay on him."

"Does' he get along well with a mop?" Maggie asked sweeping her hand in an arc in front of her "As you can see all the flooring we have around here!" She huffed and rubbed her forehead. "Lord knows I can't keep up this time of year … these kids track mud all over the place!"

Michea smiled and nodded. He was really enjoying the chili, so it wasn't really clear if he nodded because he loved chili or mopping.

Pointing with his thumb at Michea, "Ya see Maggie, he's really good at it … mopping that is. Taught him myself!" Sam looked at his watch nervously. "So what ya say woman, can the kid stay?"

"Why I never said he couldn't Sam," Maggie replied warmly. "Before you go, let me get you the home's business card … our phone number has changed."

Maggie disappeared through a doorway off of the kitchen. Sam figured she had a small office back there. He looked over to Michea and whispered across the table, "She's a terrific old broad … still got a nice looking behind!"

Maggie came around the doorway back in to the kitchen, business card in hand, "That's what Mac always said."

Sam glanced over to Michea who was stuffing his mouth with crackers and chili and not really paying too much attention to anything else, "She's got really nice hearing too!"

Maggie handed Sam the business card and a dishrag to Michea so that he could wipe his face. Sam promised he'd drop by in a few weeks after he completed the store's inventory. "Gotta keep the book keeper happy," he said just before departing for the bus stop.

In the first week to follow, Michea adjusted well, but never really settled into the group home for boys. He was restless and did things like sneaking out past Maggie's curfew rules and spending his money on bus rides around Sam's end of town. He spent hours into the night just watching the Turks hang out. They never suspected his presence, for most of the time he hung out with them, high up in a nearby tree. Even once lying atop the very bus shelter roof, they gathered under. Sometimes the cigarette smoke would curl over the top of the shelter and Michea would cover his nose. He had developed a bit of a cough lately and decided this would be his last scouting trip.

Michea would always make it back to the home in time to do the early day's chores. In between that time and noon, he'd curl up on his bed and nap. Maggie would call him out to run a few errands to the grocery store with the little red wagon. One of the home's residents, Marty would tag along. Marty was only ten years old, but looked up to Michea as his hero and role model.

"In the spring," Marty exclaimed, his eyes wide with excitement, "we'll go hunting? You'll teach me to hunt a wild boar?"

Michea looked down to Marty and smiled. Except for the blond hair and freckles, he was a lot like himself when he was young. However, he was a very accomplished hunter at Marty's age. "Must wait for the season to pass," Michea responding carefully. "Too much snow."

Marty seemed to be, as Michea observed, very smart about many things. Hunting and trapping was not one of them. So Michea tried to 'put-off' young Marty, as he knew he could not hold true to such promises of teaching him to hunt.

"I'm really good at catching frogs," Marty said, looking up to Michea as they walked along the sidewalk with the wagon rattling behind them. Michea stopped and looked down to Marty. Squinting one eye, Marty continued, "I just line up on 'em with a stick an bop'em on top the head!"

"Why you do that?"

"Brice, says it knocks the warts 'off-em', so when I go to pick 'em up they won't give me warts!"

"Brice catch many frog?" Michea asked.

"Oh boy, you bet!" Marty said excited. "You should have heard Mrs Charles scream!"

Michea nodded knowingly, then held the grocery store door for Marty. "If I'm still here in the spring, we'll go hunting."

The grocery order was ready to pick-up when Michea and Marty got there. They each loaded up the wagon with half the order, then came back and finished up the delivery to the kitchen of the Boy's Home. Marty knew of Michea's plans to visit the University and wanted to tag along. So he offered to help Michea complete the remainder of the early afternoon's chore list. At the University, Michea gave Marty some coins to buy some candy and sit in the vending lounge until he finished business. Trouble started when Marty's candy bar hung up in the machine and he tried to reach up through the vending door slot to free it up.

Marty's arm got hung up at the elbow inside the machine. That is when the Security Guard heard his screams and came to Marty's reluctant rescue.

Michea met Doctor Alan Wheeler that day. He also took flight from security and decided to evict himself from Mrs Mac's Home as more of a quarantine effort as he was beginning to feel ill. Michea knew sickness and how terrifying that was within a village much less a home for children!

Flight to the men's shelter came quickly as Professor Clark Pope provided car transportation from one place to the other.

Later that evening, everything fell apart for Michea. Coming down with pneumonia nearly killed him. Michea could not recall those emergency room moments. The only moments he recalled, were the second visit to the garden. It was a privileged visit he knew. And the more about that place and the glorious warrior that tended to the garden, the more he understood his purpose in life. So it was, out of this visit, Michea felt stronger and wiser than ever before.

He learned that there are two pathways in life. One that crumbled behind you and one that crumbled ahead of you. Together they were known as the past and the future. Someday along the time-continuum, the paths shall meet. And that will be the day the world ends. Indeed, there was something more to this, yet he did not quite understand. His visit was cut short with the great warrior before he could ask the question, as to whether they would meet on the physical side of life someday.

*****

Not meaning to make a joke of this, Michea told Alan Wheeler of his near death experience. Of how enjoyable it was to see his friend the glorious warrior. And he spoke about a lot of things, but the one thing Michea related to was his yearning to have had a longer visit.

"Michea," Alan said. "You realize from what you've told me, you would have to have another near death experience, before you'd see this glorious warrior again!"

The smile faded from Michea's face. "I am young. There will be many more."

Laughter erupted from them both. It was as if Michea could not wait for his next scrape with death. Over the past week now, Alan spent most of all his time here in the hospital, taking only a few moments to lecture his classes every other day and jotting down notes and instructions to his teaching aids. The rest of his work was done within the walls of Michea's hospital room.

Michea was very interesting to Alan. There were things about Michea that confounded Alan. Such as, Michea had spoken of his life and his growing up in the Ponctoan Settlement and how he came to where he was now. And he said all this with such conviction, that Alan wondered if Michea suffered a mild form of schizophrenia or amnesia.

Michea wasn't well, Alan knew. He was also certain Michea was a nice young man who wouldn't hurt anyone. He had a wild imagination, Alan knew, but yet he possessed an uncanny scientific tact about himself. As if he understood the workings of the universe in ways unimaginable. He told Alan that the time for him to go back to his home was near. He explained that he did not have the three weeks the doctor told him he needed for additional rest. For the following up visits, the tests, the x-rays and whatever else there was planned for him; there would be no time for him to return period! So it was a good thing that today, when the doctor smiled and said, "Michea, you're ready to go home!"

He then added, much to Michea's chagrin, "But first you will have to take one last respiration treatment. And that won't be until tomorrow morning, when the therapist makes her rounds."

Chapter 16:

Homecoming

Being born the way Michea claimed to have been, amazed Alan. Where did this young man come up with such tales? And to have arrived here from a place many centuries past, was simply too hard for Alan to take seriously. Sure, Alan had seen the arrowheads and the x-ray of an arrowhead lodged in Michea and all the other such tales, that went with those injuries, but that was something to be told about over a camp fire to a bunch of gullible kids. Not to a man of real science! Alan felt as accommodating as or even more flexible in demeanor than most men of his profession. And not to take such tales wrong and Alan didn't, for in him, there was not a fiber of malice in all the spectrum of his being. It was just now, he had lost patience in Michea ... and for a good reason.

"Michea," Alan said harshly, "all this talk about, there being no time to make your follow up appointments with the doctor, is rubbish!"

The two stared at each other for a moment. Michea looked confused. "You mad at me ... for what?"

"I'm not mad at you Michea," Alan protested. "All this talk about your people needing you. So do they need you shipped to them in a box?"

Michea just shrugged. "There is a doctor where I go. I will see him."

"Well," huffed Alan, "this place you're going … well I am going there too!"

Alan knew he was patronizing Michea. Nevertheless, to see the look on Michea's face would have been priceless, but much to Alan's surprise, Michea never flinched when he simply said, he wished Alan luck.

And yes it was known, about the special spear point. Michea had told Alan about it and that he searched for it. He also told Alan that without it, going back was impossible.

"You forgot one little thing," Alan said, "The spear point! Without it you aren't going home Michea."

"This is why," he nodded, "I must double my search to both day and night … there is no time for following up with doctor."

Alan drew a big sigh. Michea knew that Alan was getting tired of his talk about leaving. To Michea, Alan just did not understand the importance of his people's future. To Alan, Michea did not know the importance of a doctor's advice!

The solemn atmosphere of the room shifted when at the door, the familiar faces of Sam and his wife Sara came quietly in.

"Heard about my cigar-store Indian being in the woodshed!" exclaimed Sam. "So off on a bus we come!"

"Yes," added Sara, "and little Marty is in the waiting room. Shame, he's too young to come to the room here. I mean after all, it was Marty that told Mrs

Charles, who then called Sam, - that Michea landed in the hospital with pneumonia!"

Michea's eyes beamed with joy to see all his favorite people in one room together! "Marty is a good scout," Michea said.

Sam went forward and introduced himself and Sara to Alan Wheeler.

"Sam is a good teacher," Michea told Alan. "He show how to make sundaes and how to use apron!"

Alan shrugged, "How to use apron?"

Sam waved him off the subject. "You don't even want to go there," snickered Sam. "But he is an accomplished artist!" With that, Sam pulled from a shopping bag, a pack of colored pencils and a sketch pad for artists."

While Michea was wide-eyed and ripping open the pencils and sketch pad, Sara said with an impish grin on her face, "Sam, aren't you going to give him the thingy?"

"Huh, oh!" Sam exclaimed. He reached into his coat pocket and carefully extracted the Ponctoan spear point! Michea wasn't watching any of this and was busy looking down at all the beautifully colored pencils. Sam lowered the spear point to drop into Michea's lap.

Everyone in the room quickly ducked, for it rained colored pencils in the room! Michea held it up proudly. Alan's eyes widened, "I can't believe it," he said slowly, "it's true … I just can't believe it …"

"Where did you find this?" Michea said evenly.

Sam glowed proudly. "Musta fallen outa the punk's pocket." Sam made a sweeping gesture with one hand. "Got kicked under the soda bar where you landed him." Michea nodded with a smile. Then Sam continued. "You know that nagging water line that drips all the time and I gotta keep replacing it? Well, I dropped a special packing nut and it naturally, like everything else, just rolled under the damn bar. Had to get a flashlight to find it under there ... an while I was looking that big old arrowhead of yours was laying there!"

Never was there ever a bigger sigh of relief exhaled by any human being in the world. And that sigh belonged to Michea. So now it was known to Michea, that John Gravy *did* have his spear point after all! Ironically, Michea would have had it sooner had it not fallen out and ended up under the soda bar.

"Well thank goodness for leaky water lines Sam," Alan said.

"Better not lose anything else under the soda bar because I finally fixed that stubborn water line ... knock-on-wood!"

"Sam stop it!" Sara protested. "He always knocks on my head when he says that!"

"Well," Sam says bragging. "Ya don't live to be seventy-two years old and not know how to do plumbing!"

"How's gun Sam," Michea said innocently. "Shoot any enemy yet?"

Sam's eyes shifted to Sara. He was a bit taken aback by this off the wall comment. "Sam?" Sara asked turning to look at Sam's sheepish grin. "You never told me you bought a gun!"

Michea piped in, "Not barter for gun. Took away gun from enemy …"

"Michea!" Sam snapped. "Let's talk about something else."

"You bet!"

Sara collected and arranged the colored pencils back into the tray they came with. She then gave Michea a kiss on the forehead and wished him well. Sam gave him a strong handshake and assured him that all was quiet back at the store. Michea assured Sam that he would drop by the store before he left for home.

What Alan didn't know for certain was, he and Michea would do that tomorrow after being released from the hospital. Surprisingly, Michea seemed in better health than Alan expected. After leaving the hospital using the wheelchair departure to Alan's car, Michea gave directions to Charles Pharmacy and Sundries. "I know where it is Michea," Alan replied. "I have a drug tab there."

Alan barely parked the car when Michea launched out the door. He was making way down the street side of the drug store, stopping where he landed the night he fell out of the time-continuum. By the time Alan caught up with him, he had already surveyed the area for black holes with the spear point.

Michea turned to look at Alan. He shoved the spear point into Alan's hand, "Look," Michea said excitedly, "this way!"

Alan turned an eye to focus through the little bubble window in the spear point. And in the direction, Michea pointed, Alan stood aghast at the sight of a stream of tiny black holes. "Oh my God," Alan exclaimed. "They are beautiful!" Alan felt a tug on his hand.

"Come," Michea said with a tone of urgency in his voice. "We must get some of Sam's coffee!"

Sam was just putting away his boots, hat, and coat when Alan and Michea came into the store. When Sam turned to see who rattled the bell at the door, he saw Michea towing a rather dazed looking professor behind him. "Well, what a surprise," Sam mused. "and the doctor too!"

"Coffee ready?" Michea asked.

Apparently, it wasn't. Michea should have known that Sam never put on coffee until the sidewalks were scraped. Sam had just finished that chore and was going on to the next chore … making coffee.

"What's wrong with you," he asked Michea. "Yer hands broken?"

Michea knew this meant he wanted him to make the coffee. Michea loved doing this. So he looked at his hands as part of the shtick and then laughed, "You bet, Sam! Getting on it right away!"

Alan was real quiet the whole time. He just sat at the soda bar and stared, his mind far away. He knew what he saw out there. He knew that the area surrounding this stream of black holes were distinctive for this area and apparently nowhere else?

"Michea," Sam asked, pointing to Alan. "You didn't whack him over the head with something did you?"

"No, he is just spending time in his head." Michea explained. "Like you say, dreaming in the daylights"

"Day dreaming Michea," Sam corrected. "So you came to visit or stay … I figure with that gun, it'll be safe-"

"No Sam," Michea quickly replied, "came to say goodbye … going home!"

Sam thumbed a glance to Alan, and then back to Michea, "He's driving ya home?"

"Maybe Sam," Michea shrugged. "I will try to bring Doctor Wheeler with me. It will be good to try two people again."

Alan snapped out of his thoughts. "I don't understand?"

"That's okay," smiled Sam, "neither do I."

"I'm going to miss your coffee Sam," Michea said holding out his hand to Sam. "Come Doctor."

Sam waved the two off as they exited out the front door. Too bad he did not follow them out. Had he known the method by which they both departed it would have certainly grabbed his attention. Michea had Alan stand directly in front of him, then when the time was right and Michea had ripped a black hole with one hand; he crossed his arm across the chest of the Doctor and lunged them both forward and through the portal, into the time-continuum.

In the next moment, Alan stood up looking around and seeing no street, no store, just a cobblestone walkway beneath his feet, nearly the width of a single lane road. This walkway was crumbling out beneath Michea's feet, yet he seemed not to be alarmed. Alan backed away watching Michea being suspended in space. Michea seemed to be waving him to follow. "There is no walkway left below your feet!" Alan protested.

Michea looked down, "Yes there is ... come Doctor, we must hurry!"

Nevertheless, it was and neither of the two would understand that for each one, time goes forward and not backward. It was in the turn of this time-continuum along these parallels of magnetic fault lines. Alan had no time to sort this out. He knew he had slipped into a time warp of sorts and he now could see all events of the day about the town he lived. He was tempted to look forward to the future, but instead followed Michea, who was becoming impatient. "Alright," Alan replied, "I'm coming!" Trusting that where Michea went, he could too, was the last he saw of Michea. Alan stepped out over the void of space and dropped back into the spot on the sidewalk where he and Michea departed from.

Alan stood there dumbfounded, even wondering if perhaps Michea would come for him. Seemed obvious to Alan, something went wrong. Clearly, Michea had them both leap through a small black hole into space and time.

Alan wandered back inside the drug store. He took a place at the soda bar, with Sam looking on, "You look a little tattered there Doc," Sam said. "You wanna buy a comb?"

"Think I'll have another coffee,"

Sam shoved a mug in front of Alan and then topped it off with hot fresh coffee. "Say, where's Michea?"

"He's ah," Alan said slowly. "He's home."

"Would you like anything else ... cream and sugar ... maybe an ambulance or something?"

"No Sam," Alan replied looking up to Sam who looked worried. "Um, do you know where Michea is from?"

"Thought he told me he was from Milwaukee, but I figure he was from some Indian tribe that vanished a few centuries ago."

A long silence passed as Sam dropped some of Michea's pencil sketches down in front of the professor. Alan studied every one. Sam told Alan about his experiences with Michea, starting from the day they met until now. Alan listened contently to every word.

Alan then shared with Sam, his experience in the time-continuum, for which Michea made his departure back to his village obviously many centuries in the past.

"Well, I knew it!" Sam said smiling. "I learned to trust that boy with my life. He'd never lie, though at times I thought he was a little off his rocker, ya know."

A dubious grin marked the face of a weary professor snapping out of his thoughts. "I've got so much to do! I've got to get Professor Clark Pope to tell me about Michea's tribal life and I have to revise my theories ..."

"Why don't you just walk down to Glens Gun Shop," Sam offered, "just a few blocks down the street. He knows a lot about Michea's tribe. Those two had a rousing conversation about it a few weeks ago."

"Thanks I will," Alan said as he stood to head out the front door. "Oh yes Sam, make sure you have plenty of salt, saw a big ice storm heading this way!"

"What here?"

"Yes here. See you later Sam."

"Damn weather people," Sam snarled. "When are they gonna get it right … Hey Doc, how'd you know that?" Too late, Alan had already disappeared through the door for the gun store.

*****

Alan had a most intriguing conversation with Glen over the origins of the Ponctoan Indian society. He marveled over the beautifully sculpted obsidian sabre Glen had shown him. When he told Alan that Michea's mother made him this sabre and that he had killed a Calusa Warrior Prince with it, he held it for a long while, just studying it. "You know," Alan said, "*This* is the same material that the spear point was made of ..." Glen agreed.

"Excuse me if I am just thinking out loud here," Alan said, "but with his spear point, the main difference was the window in the center of it. I saw it and I looked through it. It was like nothing I have ever seen. And I tell you Glen, I have personally experienced a short jaunt through time … moreover; I saw Michea beckoning me to follow him home to his village."

Glen stared at Alan for a moment. "Look Doctor," he sighed, "I really want to believe this, but I am too much a realist. I know without a reason of doubt; this young man is a full blooded Indian. Don't know what variety or tribe mind you, but I feel he is Native American."

A smile grew across the face of Glen when he recalled Michea reacting to various bows and arrows. "He sure loved that compound bow!" Then his attention shifted back to Alan. "I don't doubt a word you've told me, but I am just saying I find it very hard to believe!"

Alan knew it was a shock. It was too big a shock to be absorbed all at once. Given a few days, Glen may find it easier to relate to and come forward with anything else Alan needed to know. But for now, an ice storm was on the horizon and headed in within the next hour. "If you don't mind closing shop a few hours early," Alan told Glen, "you'll find getting home a whole lot easier. Ice storm headed our way in about an hour!"

Glen smiled and thanked Alan for the weather advice, but opted to hang around for a package delivery.

Chapter 17:

Unfinished Business

Michea arrived back at the Settlement too late. The battle Ciabee warned of had come to pass, leaving behind death and destruction of most of the village residents. Gone were many of the faces Michea grew up with and loved. Many warriors perished in the battle. And many innocents slaughtered without regard to age or sex. It shocked and saddened Michea to learn the Chief had been savagely executed in front of his own people. The medicine Doctor, Ciabee, was found still alive with three arrows in him, clinging to life inside the Ceremonial House. And then when Michea could barely stand any more, he was informed his mother was also murdered by the Calusa Chief whom saw her as a threat when he learned she made many obsidian arrowheads and sabres that slew his warriors.

And true, with Waukee gone, much of the Ponctoan battle fire was extinguished.

As the battle went, Clugar and his Warriors did the best they could, but Calusa warriors broke through and overtook the village. Clugar lost all but eight Ponctoan Warriors. He managed to save the Doctor by dragging him off into the jungle where they could wait until dark and take back their village. Ciabee was taken back to the Ceremonial House and treated for his wounds.

There was no treatment for Michea's wounds. He slowly walked the village. No one seemed to recognize him now. Michea's clothes were of a color strange to those eyes that darted about and around trees and doorways of huts. Peculiar, were the footwear that made noise when he walked and shiny buttons down the front of his jacket. It was hot and Michea discarded these burdensome clothes. In his hut, lay a change of clothing that barely fit him now. Just like Ciabee, he thought. He looked around, instinctively looking for his mother. She met her death at the hands of a brutal murderer. And although he knew the entire village was grieving in a state of shock; he knew the understanding words of Ciabee over his tardiness in returning to the village. Yes, Ciabee had forgiven him. Although, he felt it was a different story with some of the remaining warriors; and, in particular, Clugar, who seemed very distant from Michea.

Michea spent the next two weeks helping others rebuild. In between, he stepped into the time-continuum, to check for threats from the Calusa. Things looked quiet for the next few months. Even so, talk of relocating the settlement was reverberating throughout the village. By now, Michea laid to rest his most venerable friend, Ciabee. It was a solemn occasion that went much too quietly. In these days of sadness, there would be no celebration to mark the life of this enlightened man. The entire tribe would spend the whole week in observation out of respect, for their Chief. It would be this way, however times being what they were, days passed slowly and difficultly while everyone grieved and rebuilt their homes.

Now, too many voices in the tribe objected with the idea of staying here. They claimed that the Calusa were just waiting for them to rebuild before they return to finish off the entire tribe. Michea didn't believe this, but if it was true, then it was time to plan a better defense. Michea first came forward and addressed his tribe of people, that there would be no attacks for at least a few moon phases if not more. And so, as far as Michea was concerned the matter was closed.

"You position yourself cleverly," Clugar sneered as Michea passed by. "You think you are the medicine Doctor?"

"Why," replied Michea. "Just because I told you that you done well in my absence?"

Clugar drew a deep breath. Something was bothering him. "It was the Doctor who sent you off. And when you returned, you returned too late."

"It was unfortunate," Michea said.

"You paid with your mother's life."

Michea just stared in the face of Clugar. "You tell me how she died. You tell me everything!"

Clugar didn't doubt the fire in Michea's spirit. In Michea's well composed exterior, was an explosive fire that could be ignited. If Ciabee trusted Michea with the spear point, it was obvious he would be the only hope for the Ponctoan tribe.

"I am not sure you are worthy of my time," Clugar snarled.

Michea took Clugar down with a thud, his hands gripping his throat. Clugar could not get out from Michea even if he wanted to. Clugar had fallen from a tree and injured his elbow. It had been all he could do to muster the strength to haul Ciabee into the jungle the day of the Calusa assault. Michea now, could sense this and eased his grip on Clugar's throat. "I was not there," Clugar softly confessed, "but I will take you to someone who witnessed your mother's death."

Clugar took Michea to the edge of the village where a lone hut stood. Ram-shackled in disarray, there was a sparse few arrows in sets of four leaning against a rail made of cypress. In front of the doorway, a small circle of tools lay before a white haired Indian with no teeth. He remained at work, toiling about a sheath of deer hide, used to make a quiver. Michea remembered this man. He had come to visit him a number of years ago. The arrow maker, he was and he looked up to welcome Clugar and Michea.

"Tell him," Clugar demanded of the arrow maker, "about his mother's death."

"She was very brave," the arrow maker stated. He went on to explain in gory detail of how his mother was dragged from her hut, strung up and while she begged for her life, slowly filleted and eventually disembowelled using the tools she had made.

With nothing more to say, Michea turned and left, tears streaming down his face. Two days he spent going through the family's memories and recalling all the times he spent with his mother. Then he called for Clugar to bring him his sabre. With his sabre in hand, he walked to the end of the village settlement and stopped before the hut of the arrow maker. The old arrow maker, so involved in his work, never noticed Michea walk up to him. Suddenly a small puff of dirt rose up in front of the arrow maker's knees. As the dust settled, there lay Michea's lucky arrowhead. The obsidian arrowhead his mother made special for him. The old man reached forward and picked it up. Slowly he turned it over and over again. His head nodded. "I will begin at once," the arrow maker said.

Moments later Michea met up with Clugar. "I am leaving for the *spirit-side*," Michea said. "I will be back in less than a moon phase."

Clugar nodded. "I trust you will tread carefully this time!"

Michea smiled and took up his sabre, the spear point in his other hand. Moments later, after entering the time-continuum, Michea searched for the precise entry point into Davistown. He saw an ice storm had coated the town in a glassy coating. The following week ahead looked much warmer. He could see Sam scraping away the last of the icy remnants left by the storm.

Sam had just looked up from his scraping when Michea slipped out of the time-continuum not a few feet away from him. Sam nearly shit himself. He'd never seen such a sight as this! He had never seen Michea in battle dress! There was a distinctive smell about him. A smoldering smell and as he strained a better look, he saw he was smeared in soot.

"About time you showed up," Sam said. "I see you brought your own ice scraper!"

"You bet," Michea smiled. "Looks like ice melting though?"

"Yeah," Sam said scratching his head. "Have you had your coffee this morning?"

"Is it ready?"

"No."

Michea shrugged, "Well is your hands broke?"

"Come on," Sam laughed, "you're gonna need a bath don't ya think?"

"Took a smoke bath yesterday," Michea said. Of course not everyone knew what a smoke bath was, but Michea didn't know this. Smoke baths while making you smell like smoke, kills germs on your body and in a sense makes you clean.

To Sam, Michea had only been gone a few days. But in Michea's reality, it had been nearly a month. So Sam marveled at Michea making love to his coffee. "You know," said Sam, "I outa pick you up a percolator to take home with you!"

"No electricity at home," Michea replied.

One of the things he thought was peculiar about Michea at first was his fascination with electricity. Michea had sampled electricity from the light socket of his table lamp several times before he decided that it truly existed! And Sam enjoyed calling him Einstein for having come up with the deduction that electricity makes things that require it … work!

"How many feet of electric extension cord to reach your village?" Sam asked.

"Maybe possible, maybe not," Michea said, pondering the question.

Sam was propping up the war sabre in the corner behind the soda bar. "Hey," Sam smiled saying, "you wanna have some fun?"

Michea raised his eyebrows, "You bet!"

Sam raised the receiver on the phone near the far end of the soda bar. He dialed the number to Glen's Gun Shop. "Good morning Glen, how you doing?"

"Alright I guess," Glen replied sounding tired. "Spent the last few days right here. Wanted to be here in case of a breach of security in the system you know."

"Yeah, ice storms can really mess up stuff. Say," Sam continued, changing the subject slightly, "about that Ponctoan war sabre tip, you still got it?"

"Sure do Sam and it ain't for sale"

"Sure you still have it?" Sam teased, looking at Michea's sabre leaning against the wall.

"Why wouldn't I," Glen said suspiciously.

"Better check Glen."

Glen told Sam it was there not moments ago. He had just placed it in its new showcase just a few hours ago. But when he went over to look at it, it was gone!

"Sam, it's gone! It was in the display case locked up with a key and now it's not there!"

"Glen, settle down. Come down to the drug store, I gotta surprise for you!" And with that Sam hung up. Michea barely had the time to finish his coffee, when Sam had Michea ready to pose with the sabre the moment Glen walks through the door.

"Here he comes," Sam whispered to Michea while peeping through the window display of the store.

Glen swung the front door open. There stood Michea, sabre in hand. It was a glorious weapon, adorned with an eagle feather near the base of the obsidian blade. The cypress wood pole it was mounted to was evenly grey in color and had deer skin tightly wound near the end of the handle, two hand widths wide. Glen stood there with his mouth and eyes gaping wide open. "What the ..." Is all Glen could say.

Michea reached out his sabre for Glen to examine. "This is my sabre," Michea said. "I believe years later, you acquired it. Brought it back with me so you could look at it."

Glen could barely contain his excitement. He studied the obsidian blade. "I know this stone like the back of my hand and this is the one!" Glen looked like a little boy, with a bright shiny fire truck. He shook his head, "I just can't believe it, even the tip is there!"

Michea just smiled. "I knew how much you liked it, so it is yours!"

Glen blinked his eyes a few times. Sam thought maybe Glen was about to cry. "I remembered what you said about this weapon. It served you well in battle?"

Michea proudly nodded. He knew Glen wanted to hear him tell the story again. So they gathered around the soda bar and all enjoyed a cup of coffee while Michea recounted that night time battle and how he slew the Calusa's Warrior Prince.

Glen was misty eyed over the part of Michea's new wife being shot by a muzzle loader rifle at the campfire. "Michea," he said, "you come into my store anytime you need something my friend and it's yours!"

Those were the very words Michea needed to hear right now, "You don't know how important that is to me right now Glen."

Glen shook Michea's hand and left with the sabre. Sam looked over to Michea, "You made a very good ally."

"It is so," Michea said. It was what Michea came here to do. With only eight warriors to protect his people, he needed those bows and most of all, as many modern arrows to go with them. Another important ally was Doctor Alan Wheeler. Michea knew it was important to make the Doctor believe in him. *And Sam?* He was every bit like a dad ought to be. Compassionate and smart about everyday things. Most of all he knew medicine and he had a heart bigger than the moon.

Word gets around in Davistown, especially if you are excited about one primitive Indian sabre.

Already Glen was on the phone to Alan Wheeler, that Michea was back in town! Alan was eager to see Michea and arranged to have a meeting with Michea at Glen's Gun Shop the following afternoon.

When they arrived, Alan introduced his colleague Doctor Clark Pope. Clark had a boyish look about him. Sandy colored hair and thick plastic tortoise shell eyeglass frames, which he constantly pushed up with a quick flick of his hand. He was fun to look at, Michea thought. Quite animated and when he shook hands he really shook your hand … vigorously. While shaking Glen's hand, Clark's glasses actually slipped off his face and before Clark could catch them with the notebook clutched in his other hand; Michea snapped them up in mid-air and put them on himself. Michea stood there momentarily. He looked around, sporting a goofy grin and now, all formality was dissolved with laughter.

"I can fix those frames for you Clark," Glen said. "Only take a minute … you guys pull up a chair."

Moments later, amid the rumble of conversation, Glen returned from behind the counter and handed Clark back his glasses. "Try that," he told Clark.

Clark slid his glasses on, shook his head, then looked over to Glen, "Perfect, better than new. Thank you Mr Porter!"

"This is Glen's Gun Shop, not Mr Porter's Gun Shop," Glen corrected Clark.

Clark nodded his head respectful of the motion Glen put forward. In addition, Glen continued, "I prefer we go on a first name basis."

"Exactly," said Alan.

"Makes sense to me," Clark said smiling.

To start the discussion Glen asked Clark to tell everyone what history knows about the Ponctoan People. Clark was to the point saying, "Not much Glen! That is why I am here. I hope to gain some insight on this tribe. So, Michea, help us here. Tell us everything you can about your people."

Michea started out telling them about the simple life he led growing up as a boy. He moved to telling them about all those in the tribe that made the most influence on him, of how he learned to hunt and of being trained to survive and protect his people as a warrior. He talked of the other tribes that were pushed into extinction by the Calusa tribe and how they were now persecuting his own people. Finally, as the hour passed, he told them of his latest return and the murder of his mother.

Glen, being full blooded Apache Indian himself, understood well of the persecution of war. Not that he had personally experienced such, but that he read everything he could about his heritage as an American Indian. Glen stood up and pledged, "Gentlemen," he said, "I believe we can level that playing field!"

"I have come to the conclusion," Alan added, "that we cannot go back with Michea. As you know, I tried but going back to a time before you, yourselves have been born … is simply impossible … for now that is."

Michea nodded. "It is so," he said firmly. "But Sam say it may be possible to run an electricity cord to my village so that we can enjoy coffee."

The thought of that made everyone laugh. "That Sam is something else," Glen remarked. "Extension Cord Theory!"

Alan remained silent. His eyebrows furrowed and he sat seemingly in deep thought. "No," Alan said loudly, "Sam could very well be a genius!"

"No, he would be an Einstein," reminded Michea.

"Exactly!" exclaimed Alan. "Better yet, a rope traversing time and space … could it work?"

Sam was confused. "What are you nimrods talking about? A rope ain't gonna send electricity to nowhere … might as well just use a damn string!"

"Rope, string what difference does this theory make if it defies explanation," Clark added.

"Michea," Glen asked, "if you enter that portal thing with a rope, it may be possible that we could follow you into time."

"Maybe so," Michea replied. "I would like to try that!"

*****

That evening, Michea, Alan, Clark, and Glen along with Sam set up a place next to the drug store where Michea said was the best spot. Glen held a coil of rope in his hand, the other end of the rope; Michea held, as he narrowed in on a black hole using his spear point. All marvelled to see Michea being swallowed up into nothingness and the rope … it reeled out from Glen's hand as if an invisible fish had taken it. Clark passed his hand behind the point of where the rope disappeared. "Wow," Clark mused, "this is weird!" Glen was gently pulling on the rope.

"Michea is still holding on to the rope," Glen reported, with an amazed look on his face. Looking at Clark now, Glen asked him to take the coil of rope. Clark complied and took the rope while Glen gingerly slipped his hands along the rope to the place where the rope disappeared into the time-continuum. For a moment everyone could see Glen's hand disappear, "I can feel the rope on the other side!" Then suddenly, Glen appeared to flatten out as if he was only a picture and in a split of an instant, he was sucked into the time-continuum. Alan offered to be next, then Clark.

Sam was alone now. "Well," he muttered under his breath while looking at the rope. "You don't live to be seventy-two years old just to pass up an opportunity like this!"

"Michea," Sam yelled as he appeared on the stone walkway with everyone else. "You better reel that rope in off the sidewalk." Sam was now looking down at his store and the rope lying on the sidewalk. "Someone will come along and trip on that … don't wanna get sued you know!"

Michea obediently pulled the rope hand over hand until he stopped. "Someone is pulling back on the rope," Michea said looking confused. Sam carefully peered over the edge of the stone walkway. "Dammit," Sam cursed. "It's that punk, John Gravy! Go ahead Michea pull!"

John Gravy was really doing his best at this mystical tug 'o war game. That is, being laid up with a mechanical knee brace and all. It was all too clear to Sam and the others that Johnnie Gravy was close to being pulled through the portal of time. Sam looked at Michea, "Give it a sharp yank! He'll let go!"

Michea did as Sam ordered. But John managed to hold on. And as he did, all peered down engrossed in the action as one watching an Olympic sporting event.

Now, standing on the walkway of time, was Alan, Clark, Glen, Sam, Michea and a reluctant street punk named John Gravy.

John, confused and shaken by the sudden shift in his surroundings, just looked around wide eyed. He saw Sam's store below him and his street gang members milling about just as confused as he was. "Hey," he said looking at Sam, "what the hell is going on here?"

Obviously, Michea didn't want to have to put up with this punk again and grabbed John by the arm and swiftly slung him out and away from the walkway—on the other side! Clark was peering over to see John Gravy land into what appeared to be an old cemetery. "He went somewhere else," Clark observed. "Sure don't look like anywhere in Davistown!"

Glen admired Michea for his swift action. "Well, that's taking out the garbage if I ever saw it," he chuckled.

Everyone wanted to stay a while longer, but Sam was worried, "Gentleman, I have to get back to my store!" And no one could really blame him, seeing the hoodlums milling about just around the corner from his front door. And no one wanted to see Sam go alone, so they all stepped off the edge and onto the reality of the city sidewalk. As they appeared, the hoodlums approached Alan with the question of the whereabouts of Johnnie Gravy. Then it was seen by the gang, Michea appearing before them out of the night, rope in hand. Seeing this, the gang scattered without a word, taking flight down an alleyway.

*****

The hand that reached down to give Johnnie Gravy a hand up was unfamiliar in that he had never seen this guy before. "And who the hell are you, *you asshole!"*

"The name's Doctor Maxwell Tanker," the portly figure of a man said.

Chapter 18:

The Teacher of Lessons

To all those in the village that had seen Michea arrive back home, no one but Clugar could see the change in him. A spark of wisdom that was hard to define in Michea's eyes shone stronger with each trip to the *spirit-side*. And this time, Michea came bearing gifts! As he strode through the village toward the ceremonial house, Clugar could see Michea toting an unusually large quiver shaped bag. It was slung over his shoulder and with it, an assortment of bows across his chest.

Inside the ceremonial house, Michea began unloading his wares, laying each bow on the table before Clugar's curious glare. Michea looked at Clugar's radiant smile, as fistful, after fistful of aluminum hollow shaft arrows were laid out on the table; some with narrow tips, others with deadly sharp broad-tips. Clugar moved around the table like a child in a toy store. He couldn't wait to test one of the recurve bows. He rubbed the bow gently in his hands while Michea explained to him that it was made of a type of wood called 'fiberglass'. It had a balanced strength, which was even and smooth. Michea handed

Clugar an arrow. He studied it closely and remarked that it weighed 'nothing'. "For such and arrow," Clugar explained, "is not good! Shatter on impact!"

Michea laughed. "Shoot that tree!"

Clugar stepped out of the doorway and took aim on a huge cypress tree across the field a good hundred yards away. "Arrow never make it that far," Clugar sneered. Clugar was an excellent archer. He could put an arrow through a turkey at such a distance it was said. Michea would take Clugar's advice on matters such as bows and arrows, but this time he would be wrong. Clugar let the arrow fly with a distinctive snap. A ringing sound could be heard as the arrow drilled into the tree with such force. It could be heard all the way back to Clugar's ears.

"Feathers never touched cheek!" Clugar said dumbfounded.

As they walked closer to the tree, Clugar nearly tripped over his feet. He was admiring the bow, his eyes mesmerized by its graceful appearance. "We will see what this bow can do with real arrows!" But when they reached the huge cypress tree, Clugar looked at Michea with disbelief. He reached to pull the arrow free of the tree, but found it had burrowed deeper than he expected. Clearly Clugar was impressed, but Michea could see by the way he was gingerly handling the arrow, that he was afraid it would break if he got too rough with it.

"Here," Michea said, reaching up and grabbing the arrow while wiggling it back and forth so roughly that Clugar grimaced. Splintering tree bark fell to the ground as Michea freed the arrow from the cypress tree. Handing it back to Clugar, Michea explained, "It's made out a thing called aluminum. It's light and much stronger than wood arrows."

Michea was happy to see Clugar so delighted. He held his fist to his chest and sheepishly confessed his heart was pounding. He confessed not even a wood arrow would have performed this well. "It was like an old arrow," Clugar told Michea with a newfound sense of pride. "But now, I see myself like this new bow and arrow ... we will survive!"

Clugar was so into his new toy, it was all Michea could do to field all the answers to his questions. To think these arrows would last many shootings really impressed him. "And these arrows," he told Michea, "come from the gods!"

"No," Michea corrected. "Glen ordered them from Michigan." Michea shrugged his shoulders when he said that because he wasn't sure what it all meant. Then of course he had to explain all about Glen and his gun shop.

As they drew close to the ceremonial house, Clugar stopped and slapped Michea on his shoulder. "I must meet Glen ... you must bring him here!" Clugar was beaming with a silly grin.

"What?"

Clugar proudly said, "I knew such superior hunting and battle weapons would rule in the domain of an Indian Chief!"

"Yes Clugar, Glen is a proud Indian. And he begged to come here, but it was not possible. There are laws that forbid him to cross over to this place.

But you could come with me and meet him someday."

"There must be promise in your words," Clugar said, his voice deep and serious.

Michea pointed to a coil of rope on the table inside the ceremonial house, "Keep safe this. It will insure your wish will come!"

There were many things to show Clugar, but the moment to do so was interrupted by an old arrow maker standing obediently at the door. He had a 'special order' for Michea. And Michea asked him to come in and lay the arrow on the table. Michea reached into his duffel bag and pulled out a small silver metal shape. Michea flipped the top off with his thumb with a click. Then with the snap of his thumb there was a bright flame! Slightly startled, the old man settled down with a sigh of amazement in his face. Clugar dropped one of the brightly colored arrows he'd been looking at back to the table and moved closer to the flame. He ran a finger through the flame, then looked at Michea and laughed. Michea snapped the lighter closed then opened it. Flame was gone! He then re-struck the flint wheel and a new flame reappeared. "Zippo!" Michea said. The old man took the lighter and operated it perfectly the first time. He was pleased with himself and the magic fire box.

"Zippo," he said nodding his head. Michea knew he had struck a deal on the arrow and he nodded his head. Back to his hut the old man danced with his new toy.

Michea turned to find Clugar already checking out the other bows. He had picked up the Compound Bow and studied it with stymied look on his face. Michea allowed Clugar to hold it until Clugar held it up to pull back on the string. "Put it down," Michea said sternly. "That one is mine!" An unspoken code among warriors was the respect for personal property. A level higher on the respect ladder was never handling another warrior's weapon without permission, unless of course it was in an emergency.

Clugar knew right then and there that this ugly bow must possess powers far greater to even the ones made of fiberglass. Michea picked up the Compound Bow and explained that it was the only one of its kind. This single bow, he told Clugar, had the combined punch of all the bows he saw on the table before him!

While all this was sinking in to Clugar's mind, his eyes followed Michea as he held the Compound Bow. In his hand an arrow with an obsidian arrowhead. Michea slowly nocked the primitive arrow into this weapon made for the gods!

Today Clugar had been acquainted with the future in weapons and already looked confused by the sight of this arrow/bow combination.

Michea was merely checking the length of the arrow to ensure it was compatible with the Compound Bow. Too short was equally as bad as being too long, he knew. Even so, the primitive arrow was slightly longer when measured up against a modern arrow. Michea impressed, whispered reverently; the arrow maker was truly a skilled man.

Even though it was still early in the day, it was late day for Michea. He told Clugar he would not be disturbed for all of the next day. Then he would be gone by early the day after. Then he asked what phase was the moon.

"New moon in three days," Clugar said.

This was good news to Michea as he was feeling weak and tired. It was a hectic week for him. Still recovering from pneumonia, he'd been warned not to wear himself out and to return to the doctor in three weeks for a follow up. If all his friends in Davistown had their way, Michea would be confined to a bed and not allowed to do anything for himself. As weary as Michea was feeling right now, having a few nights to regain his strength before the night becomes completely black was the perfect medicine he could get.

Back to his old home, Michea made it to lie in his mother's bed. The sweet smell of lilacs and honeysuckle filled him with the sense that his mother was still about. He lay curled up on his side and drifted off to sleep. Sometime in the middle of the night, Michea woke. He was cold and he knew he needed to get up and find a blanket. His eyes still closed, he instinctively reached out like he had many times in the past. He never had to reach far when his hand fell upon a neatly folded blanket. Michea jarred awake with the thought that he had not noticed the blanket being there when he lay down. Even so, there it was and again, as he rolled himself in the blanket, the sweet smell of his mother was there with him. And now he felt cradled in her arms. Her warm and soft breath was on him as she caressed his face and told him things he struggled to understand. His heart sang of the beauty of her touch and her tender kisses on his face. When he shivered from fright, she fetched him up and held him tight and told him everything would be alright. And when he shivered from the cold, she held him close to her and kept him warm. She was his 'mum' and he was her little man!

The following day the sun rose and an excited group of warriors came to greet Michea eager to hear his stories of the spirit side. There were many questions to ask Michea about the marvellous bows and arrows. Clugar held them off, ordering them to be quiet. Gingerly, Clugar peered inside the door of Michea's hut finding him still fast asleep. Michea was not known to sleep this long into the early part of the day. However, Clugar seemed to understand Michea would sleep a long time.

And that Michea did. He slept all the way into the night. Sleeping until suddenly he awoke. He felt as if he was about to be dragged out of bed. Whenever he confined himself to his bed and pouted, his mother would always drag Michea out of bed. He thought he could hear her say, "Michea, you will get up and do your chores, even if I have to paddle your behind with a beaver tail!" Michea felt the beaver tail a few times and knew learning to do chores was a better alternative.

"Yes mum," Michea whispered. Tears streamed down Michea's cheeks now. "I love you."

With a deep breath and a sigh, Michea lay back down and wrapped himself in his blanket. And now it seemed to Michea, that his mother's presence was upon him for the sake of tending to his recent illness. And then as now, she was rocking him to sleep as she always had in the past when he was a sick or injured little boy.

Michea could hear his mother whisper, "Tomorrow you will be better."

Light shone through the doorway as the new day's sun crested the cypress bulrushes that surrounded the settlement. Water ran crystal clear from the bubbling water pit aquifer, lined in a spongy limestone bed just a few inches below the soil. From there the spring spread its swampy sprawl feeding the cypress bulrushes that thrived there. Michea started his new day, washing the sleep from his face and drinking his fill from the small creek that flowed from this bulrush. The stream ran behind his and many other huts along its way, to where it eventually emptied into the head of the Peace River. This was a rich and beautiful land. And Michea knew it was worth defending it to the last drop of Ponctoan blood, if that is what it would take to keep it from Calusa control.

Fully awake and feeling better than he had in months, Michea re-entered his hut and went about his chores of packing a duffel bag for the long hike to the Calusa village. It would be an all day hike and by the time this days sun set, he would be prowling the Calusa settlement. It would be too dark for them to see him; but not unlike a panther in the night, he would see them quite well. He checked the battery pack like he had been trained to do. And the belt and holsters and all the pouches with spare clips, were checked and assembled. The night vision goggles were a new technology Glen wanted to test them himself, but decided Michea could do that well in real battle. Michea was quite impressed and knew battery life could be as short as one hour, so a spare battery was included in a leather pack he wore on his belt. Finally, he unrolled a full set of jungle battle fatigues complete with boots. He used a small signal mirror to apply camouflage paint to his neck and face. Michea then packed his pockets full of battle goodies, strapped on his utility belt and donned a camouflaged boony hat. Before he left the hut, he slipped his Compound bow and quiver over his head and adjusted them in place across his chest. The new arrow with his lucky obsidian arrowhead took its place with four modern arrows in his quiver.

Michea slipped out of the settlement without being seen, except for Clugar who kept a watchful eye. As from the path Michea chose could only mean one thing. Fear gripped Clugar's heart. Michea was going to enter the camp of their enemy on a night with no moonlight? A night the local Indians called a 'Panther Night'.

******

Michea setup a day camp less than a mile from the heart of the Calusa settlement. Bordering the settlement on the west was the open bay to the Gulf of Mexico. To the east the jungle forestland to which Michea resided. Michea realized from his battle training over the last week, that this was a very unwise geographical location. To have no place to retreat other than the sea was a huge mistake. The Ponctoan settlement was perfect as any battle ready camp could be, especially during a Panther Night. To surround his village would require a battle force of hundreds of warriors. The Calusa had vast number of warriors to do the job, Michea knew. They had already shown this strength while he was away.

In all his thoughts, Michea worked getting ready his 'show' for the Calusa Chief. There were two special arrows, he and Glen devised and tested. It was designed on the take of the old 'flaming arrow'. Glen knew a flaming arrow would blind Michea's sight while he prepared to take aim on his target. What Michea needed was an arrow that ignited on impact. So Glen devised an arrow fitted with a golf ball jacket packed with primer cord and a few primer caps with snap pins installed. The arrow on impact would cause the snap pins to strike the C explosive caps setting off the primer cord waded inside the golf ball cover. Overall, the arrow was light, yet very explosive. Safe and easy to tote, these two arrows would open the show!

Michea shared the last of his beef jerky with a raccoon before setting out for the short hike to the enemy settlement. He re-sheathed his combat knife and set his night-vision goggles on his head; ready now, to drop them in place as he approached the settlement. Out through a clearing, Michea reached the beach, just north of the settlement. The sun was already down for over thirty minutes and now it was so dark Michea could not see his hand directly in front of his face. With even the stars shrouded by clouds; it was the perfect of all Panther Nights. He flipped the goggles down over his eyes and turned the switch activating two small CRT viewfinders in the viewer cups over his eyes. They were bright and he rolled the little adjustment thumb wheel to dim them down to a comfortable viewfinder level. He smiled. It was like daylight now.

It was low tide and Michea approached from the water's edge now. The sand was tightly packed and easy to tread on. Close to the tree line where sand dunes rose here and there was the first of many huts. He walked right by them seeing eyeballs flash in the huts and the bright flashes of a small camp fire in front of one hut. Michea slowly nocked a fire arrow in his Compound Bow. Looking farther down, he could see the Chief's hut and the ceremonial hut next door. And then he saw two horses! They were tied up outside the following hut next to the ceremonial hut. Those were the animals belonging to the Spaniard people and they had guns! At first, Michea felt a twinge of fear and defeat until he realized these rifles they used were primitive and could simply shoot one round at a time. Reloading after each shot took long moments of preparations.

The first target for his fire arrow would go to the Spaniards hut. Michea took aim and let go the fire arrow. The arrow crossed the nearly hundred yard distance in a split second. The ball of primer cord exploded in a white hot ball of energy that literally ripped the roof clean off the hut! Yellow flames immediately erupted and flumes of black smoke bellowed from the hut. It was too much light for Michea to see, so he flipped off his goggles. He as well as others could see rather well thanks to the hut fire. Two Spaniards ran for the sea to cool off in the water. No one saw Michea or noticed him in the fringe of the darkness where he stood. However, from where he stood, he could see everything going on around the village. Two men left their huts and ran right past him!

Michea drew a deep steady breath. Standing outside his doorway was the Chief! Michea's hands trembled. He had never killed a Chief. Then he realized this was not a Chief. It was not even a man … it was the animal that took his mother away! Michea clenched his teeth in anger, his eyes narrowed readying for the focused aim he would draw on this worm of a man. Reaching back to his quiver of arrows, Michea slowly extracted the ancient arrow of his people. The one with the specially crafted obsidian arrow-tip his mother made him when he was a child. Nervously, Michea nocked the arrow in place along the string of the compound bow. Pulling back the arrow, he felt the bow signal back to him that it was ready. The time had come and Michea let the string slide off his fingers. The feathers whipped his cheek raw as it passed with the speed of a lightning bolt.

Michea quickly slipped his bow across his chest as he strut to meet the pig that took his mum's life.

The Chief had slipped to his knees, resting his buttocks on the heels of his feet. Michea's broad tip obsidian arrow had passed through the man's back and protruded out of the front of his chest by a length of at least four inches. He just hung his head looking down at the awesome size of the black arrowhead that looked as if it was staring back at him waiting for him to pass on to the animal side.

Now Michea stood before him. "You will pass to the life of the snake for which you are," Michea told him. Then he grabbed the arrow and snapped it off in his hand. Gripping it like a knife he said further, "This is the arrowhead my mother made me. You killed her because of it …"

As a dishonor to the Chief he lifted his head by the butch of his hair and gouged out both of his eyes with his mother's arrowhead. "You have been sold into the next life, blind as the snake for which, you will become!"

Michea was aware of the three men running up to meet him. He dropped his lucky arrowhead into his quiver, unsnapped his sidearm holster, lifted out his military .45 automatic, cycled the slide to chamber the first round, then levelled the barrel into the face of the first man and shot him point blank. The second man no wiser than the first took the next round in the belly … and then the third man, a round in the belly as well.

Michea began to back pedal into the darkness. There was a group of warriors already assembling near the Spaniard hut.

Replacing the goggles over his face, Michea ran for the forest path. There would be no way they could follow him once he disappeared into the forest. He holstered his weapon and snapped free a hand grenade from the shackle of his vest. He ducked down in the sea oats that crested a sand dune and waited for the group of warriors to follow up the beach toward him. Quickly, he pulled the pin and held the arming spoon tight in his hand, ready to throw when the time came. Closer they came and now; Michea could see they came arrow ready and nocked for business. There were five of them and Michea let fly the grenade. He flattened himself out on the sand and raised his head just in time to see a huge orange flash of light and a deafening explosion. He didn't wait to see the outcome. He just jumped to his feet and ran into the forest. In the dense cover, he slipped down into the thicket and waited. Now the crickets began to resume their night time chorus and Michea knew he'd gotten away! His night vision goggles began to flicker a few times. Michea quickly swapped to the fresh spare batteries he'd brought along. Then knowing he had about an hour, he ran back toward home as fast as he could. The goggles were still working fine, but he knew he had put a safe distance between him and the Calusa settlement now. So he settled down for the rest of the night. Tomorrow he would move at first light.

Chapter 19:

Reflections

The light of the morning came and it glistened over a semi glassy sea. Small waves rolled to shore; rolling back and forth along with them, the lifeless body of one Spanish Explorer. The second Spanish Explorer lay on his back inside the ceremonial hut where the Chief was laying in state. An old woman cautiously dabbed cool water and aloe on the Spaniard's burned body. He moaned constantly and then finally screamed for her to leave him alone to die.

In the corner shadow of this room stood the medicine Chief. He stepped out of the shadow to welcome, his yet to be appointed Prince of Warriors. His name was Coure and he had seen his successor fall dead before him in the darkness of the night. Coure had been slow to arrive at the site of where the fireball had consumed four of his warrior comrades. Nonetheless, he heard the words of the last survivor in the grenade attack. He told Coure it was a revenge attack by the gods who were angry at having nearly decimated the Ponctoan people. The medicine chief listened to every word Coure told him. He frowned and after a long silence, he turned to the table where the Chief lay. Coure saw the gaping wound in the Chief's chest. The medicine chief slowly shook his head in disbelief. "The Chief," he told Coure, "died only a short while ago!"

Coure shuddered at this and the horrible way the Spaniard wreathed in pain. "The Chief died a brave man!"

There was a troubled look on the medicine chief's face and Coure just stared, bewildered for a moment. "Of what you told me of the god's revenge," the medicine chief said slowly reaching to remove the cloth covering the Chief's face, "it was surely so."

Startled by his own gasp, Coure threw his hand over his mouth to contain himself. He saw the face of his Chief; a face frozen with the grimace of one stricken with disgrace. The words of the medicine chief went past Coure's ears as he was locked in shock of the sight. Finally, the hand of the medicine chief on his shoulder broke Coure out of the shaken state he was in. "I know you will not speak a word of this disgrace," the medicine chief said. Coure slowly nodded.

"We will grieve this great man?" Coure asked cautiously.

"Yes," replied the medicine chief. The medicine chief turned his back to hide his expression of fear. "And we will in turn seek justice!"

Coure was unfamiliar with the difference of justice and revenge in this case. "We will challenge the gods?" Coure choked with his question as he knew why he was being sent here for.

"If you are not up to the task, tell me now!"

"The men are shaken, I will need time to sort out and devise a war plan," is all Coure could muster in the shock of such a request.

"After the Chief is celebrated and you have had your celebration for rising to the position of Prince of Warriors. Following that, we will talk war!"

Coure bowed respectfully and left the ceremonial house. The medicine chief remained. He had heard the wrenching of hearts in the village and the cries of children. Most of all he had heard the terrifying threats of the Spaniard in the room with him. Out of his pain he warned there will be many more of his kind to descend on his village. They would take charge of his people if he failed in taking up arms against the Ponctoan settlement. The bitter last words of the Chief would be enough to perpetrate enough hatred among the people. The medicine chief wasn't sure how this horrible event occurred with such spiritual like thunder and magic, but he was certain it would happen again. The Ponctoan warriors could see in total blackness and that frightened the medicine chief. They had the eyes of a panther at night and just like the Chief had described before losing his eyes, the face he saw was that of a panther!

The medicine chief's heart beat hard in his chest. His head flooded with thoughts, wonders and fears …

The Calusa from this day forward would live in fear of the Panther Night.

Epilogue:

No time like the present

Clugar had seen the bright lights the night before. It lit up a Panther Night over the Calusa village, he knew. And the clap of thunder rolled from there, all the way through the forest to the Ponctoan Settlement. Clugar had his men set up a watch last night, in case there might be renegade action on the village. There was a better chance of the sky opening up and dropping virgin womenfolk on the men of the village, Clugar knew, but to be safe was all about the ways of survival. Panther Nights were generally nights of comfort and security. Knowing that even your most feared enemy would not move on you in such darkness, unless your enemy was … Michea.

And Clugar's attention drew on a figure emerging from the forest. He was a darkly camouflaged man that Clugar barely recognized as his friend Michea!

As he approached, Clugar was mesmerized in his confident stride and the strange clothing that he wore. From the bonnet on his head to the forest green patina of the forestland in his clothing; his boots to his strangely wicked looking bow adorned him with a fearful sight that this warrior was not to be trifled with. It was not right for the likes of a Prince of Warriors to tag along behind any warrior. Clugar fought the urge and merely addressed Michea well, as he passed by on his way to his hut.

Moments later Michea emerged from his hut, dressed in the clothing of the *spirit-side*. His red and black flannel shirt and thick vest told Clugar he'd be gone awhile … maybe not. But it was certain Michea was off to the spirit side to pay homage for his successful attack on the Calusa Tribe.

Clugar watched from the center of the village circle. Michea just stood for a short moment staring at Clugar. Then, with a quick pass of his hand, he yelled to Clugar to come to the door of his hut where he was standing.

"Here," Michea said, "put this on."

Clugar looked at the thick coat Michea offered him. He put it on and it felt warm, yet very light. And then when Michea passed him a pair of boots, Clugar was astonished with the funny way it made his feet feel. "No," Michea said, "Put these socks on first!"

Before long, Clugar looked like a craggy long haired street bum. Michea grabbed the rope and gave Clugar a short lesson on what was expected of him. Michea had all the confidence in Clugar. Under all the savage roughness of this man was an intellectual.

There was one thing that troubled Clugar. And that was exactly what to expect on the 'other side', so Clugar snapped-up a leather thong he'd been toting which held a tightly bound breakfast squirrel to it and tied it to a bearskin sash he wore around his waist.

"We're going to Glen's Gun Shop!" Michea said cheerfully. "You're going to love that place!" With that, Michea snatched-up the rope and the match-lock musket previously owned by the Calusa Warrior Prince he'd slain earlier and marched to the place next to the Ceremonial House where the tiny black holes danced. With the rope and musket in one hand and the Ponctoan Spear in the other, Michea found one of the small black holes in a perfect position that suited his escape into the time continuum. Clugar watched intently as Michea dropped the rope coil to the ground while still holding on to the end of it. Immediately following, Michea swiped at something before him and Clugar felt a blast of cold air move past them.

Michea disappeared, leaving behind a rope that dangled from a place in the middle of nowhere. Clugar took the rope and instinctively closed his eyes, giving it a sharp tug to signal to Michea he was ready.

In the snap of a turtle's jaw, time had swallowed Clugar, leaving behind a rope that sharply began to reel into the continuum by the hand of Michea.

*****

Life would never be the same in the Ponctoan Village. Rebuilding the settlement and re-establishing a normal climate among the people once more will be challenging. Michea and Clugar had much to do to balance all of this; and yet, there was the matter of rescuing Michea's young bride Lichea from the place he'd seen her in the *spirit-side*.

And then, there were the evil forces which would devise plans to take the Ponctoan Spear from the people who greatly depended upon it. And for those who harbor malice in their hearts, it was one man who could transverse both the future and the past they would have to face …and that was Michea.

~Finis`~?

About the author:

Aside from his love of storytelling, T.A. Walters enjoys sailing coastal Florida waters and relaxing while reading science fiction novels. He also loves to write short story contemporary fiction inspired by his love of the outdoors and sailing the seas.

T.A. Walters short story collection is the 'Island Adventures' series include stories aboard the Southern Bristol and the Southern Fox.
All available on Kindle.

Twitter the Author:

@T_A_Walters

Email the Author at:

T.A.Walters@hotmail.com